WE USED TO CALL THEM SWAMPS . . .

I'm not quite sure when it happened. But somewhere along the line, the local swamp became a "wetland." The widespread use of the term snuck up on us, becoming current without much notice. With the upgraded name came a better understanding of the ecological importance of the varied bits of landscape we call wetlands.

It used to be fairly easy to decide what to do with these wetlands, these bogs, swamps and marshes. With some money and effort, they could be made into productive land, useful for farming or building sites, roadways or marinas. Now that we know more about the benefits that wetlands offer in their natural state, it's much more difficult to unequivocally support altering or destroying them. Conversions still happen; but the decision to make them is not as straightforward as it once was.

The question of what to do with wetlands has become a public issue. It's not clear to what extent decisions about wetlands should be made by government and to what extent these decisions should be left to private landowners.

It is also unclear how we should weigh the benefits of natural wetlands against the benefits of their conversion. What weight should we give functions that wetlands perform naturally, such as providing wildlife habitat, controlling floods, or renewing moisture in the atmosphere? How should we factor in aesthetic values and the importance of keeping ecological systems intact for future generations? And what weight do we give to other potential uses for the land and water that combine to make wetlands?

The current question of what to do with wetlands, and which wetlands should fall under government control, cannot be answered simply by consulting experts. Nor can it be answered once and for all by coming up with a scientific definition of wetlands. It is a complex political debate that needs public input.

Environmental Issues Forums on what to do with wetlands challenge participants to consider a number of different approaches to resolving wetland issues. In exploring the pros and cons of each approach, people in your community can learn more about what makes wetlands the focus of political controversy and begin to create a shared understanding of what the problems and solutions might be.

This shared sense is the seed of a public consensus that can guide our decisions about wetlands in a way that reflects what we care about most. This book, like the others in the Environmental Issues Forums series, is a blueprint for discussions that can provide momentum and substance to the resolution of this important environmental issue.

Lori Mann
President, NAAEE

The books in the Environmental Issues Forums (EIF) series are prepared by the North American Association for Environmental Education in cooperation with the Kettering Foundation. These materials are designed to help local organizations and schools conduct public meetings and study circles addressing difficult environmental issues.

Writer—Michele Archie
Copy Editor—Ellen Lambeth
Designer and Publisher—Characters, Inc.
EIF Production Team—Ed McCrea, Keith Roeth, Steven Schiff,
  Andrea Shotkin, Bora Simmons
Cover Design—Bob Hora

This second book of the Environmental Issues Forums series was an opportunity for us to test our wings, as it were. We are grateful to the staff of the Kettering Foundation—particularly to Patrick Scully, Bob Kingston and Gina Paget—for offering just the right amount of support and guidance as we put into practice what we learned in writing the first book. We would also like to thank David Mathews, President of the Kettering Foundation, for his continued encouragement.

Our appreciation goes out to all those whose cooperation made this book possible. The list of people who generously provided insight, direction, and written information is too long to detail here. Special thanks to the people who reviewed the manuscript: Ken Cox of the Canadian Wetlands Conservation Task Force and Larry Mason of the U.S. Fish and Wildlife Service.

This book was written with the support of the **Kettering Foundation.**

Groups interested in using the EIF materials and adapting the forum approach as part of their own programs can contact the EIF Coordinator at NAAEE, Suite 400, 1255 23rd Street NW, Washington, DC 20037. Phone (202) 467-8753. Fax (202) 862-1947. Additional copies of this issue book can be obtained for $4.98 (U.S.) each from the NAAEE Publications and Member Services Office, P.O. Box 400, Troy, OH 45373. Phone and fax (513) 676-2514. Copies of the first book in the series—*The Solid Waste Mess: What Should We Do with the Garbage?*—are also available. There is a discounted price for orders of more than ten books.

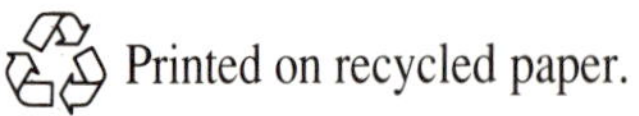 Printed on recycled paper.

TABLE OF CONTENTS
The Wetlands Issue: What Should We Do with Our Bogs, Swamps and Marshes?

WHAT ARE WETLANDS?

From Florida's Everglades to the southern shore of Canada's Hudson Bay, from the pothole country of the prairie states and provinces to the Pacific coastal marshes of Vancouver Island and Oregon, North America is dotted with a vast variety of wetlands.

We know wetlands better as bogs, swamps, marshes, fens, wet meadows, wet tundra, ponds, peatlands, muskegs and bottomlands. Generally speaking, wetlands are areas that are wet or covered by shallow water for at least part of the year. Their soil conditions, and the plant and animal life they support, are determined mainly by the presence of water.

Wetlands can be thought of as "transitional" zones, where the characteristics of aquatic and dry land environments mix. They are often found along the banks of lakes, ocean shorelines or river flood plains where periodic high water or tides provide the water that sustains them. Other wetlands exist apart from these obvious sources of surface water. They may be fed by water that is just beneath the surface, by rainfall or by seasonal melting.

Both freshwater and saltwater wetlands are subject to fluctuating water levels. Most often, springtime high water levels determine the limits of freshwater wetlands, which may later dry up because

Millions of "pothole" wetlands dot the prairies of the United States and Canada.

Tim McCabe/Soil Conservation Service

Edward Lyles/U.S. Fish and Wildlife Service

Forested wetlands often seem dry during low water.

Gene Whitaker/U.S. Fish and Wildlife Service

Wetlands are home to many plant and animal species . . .

Glenn D. Chambers/Ducks Unlimited, Inc.

And provide important habitat for migrating, nesting and wintering waterfowl.

Tim McCabe/Soil Conservation Service

Cypress swamps are found in coastal areas of the southern United States.

of summer droughts, evaporation or plant growth. Forested wetlands may even appear to have dry ground for much of the year. Saltwater wetlands, dependent upon ocean tides, may dry up as often as twice a day, or may be flooded only during occasional high tides.

Not all wetlands are natural. Human development has resulted in the sometimes unintentional formation of new wetland areas. Many of these surround reservoirs created for hydroelectric generation, flood control or irrigation. Rice paddies, fish ponds and the excavated areas associated with road construction or mining are among the most obvious examples of artificially made wetlands. Other wetlands are created for landscaping purposes or to support waterfowl.

Lisa Wilcox

Glenn D. Chambers/Ducks Unlimited, Inc.

Wetlands support a variety of recreational and commercial activities.

Tim McCabe/Soil Conservation Service

Tidal salt marshes are found behind coastal barrier beaches, or protected bays.

U.S. Fish and Wildlife Service

Some wetlands, like this stream bed, are occasionally flooded.

WHAT'S BEHIND THE WETLAND DEBATE?

Like other jurisdictions across the United States and Canada, and around the world, the government of British Columbia has some tough decisions to make about wetlands.

The immediate question for British Columbia involves the land that hugs the shore along the last natural stretch of the Columbia River. For most of its 1,240-mile (2,000-kilometer) length, the Columbia River is host to a series of dams and reservoirs that harness the Columbia's once-raw power for hydroelectric generation. But from its source near the small British Columbian town called Canal Flats to the head of Mica Reservoir some 112 miles (180 kilometers) downstream, the river runs free.

Along this free-flowing stretch, the Columbia River gives life to what writers Kevin Van Tighem and Betty Baird (as published in *Canadian Geographic*) describe as "a 26,000 hectare [64,246-acre] mosaic of marshes and sedge meadows, cottonwoods and willows, shrub thickets, ponds, swamps and running water."

Almost all of these wetlands are provincial Crown lands. That is, they fall under the control of the British Columbian provincial government. Until recently, the government had restricted development on much of this flood plain to avoid conflicts over a new hydropower project that was in the works. That project would have involved flooding some of the wetlands and filling in others. But the plan turned out to be economically unworkable, so the government has since removed the restrictions.

Now, the question of what to do with these wetlands is wide open. In trying to answer that question, the government faces some difficult choices.

On one hand, many people would welcome the chance to use this land to bolster the local economy. Along this 112-mile (180-kilometer) stretch, the Columbia River runs through a narrow valley known as the Rocky Mountain Trench. The wetlands in the Trench are often the only flat land available for development. Some people are making plans to build subdivisions or to build resorts and recreational facilities such as golf courses

to encourage a growing tourist trade. Others want to use the wetlands for agriculture. Since the government controls so much of this wetland area, many of these proposals hinge on gaining permission to develop Crown land.

For others, though, a little bit of development goes a long way. As Van Tighem and Baird write, "[Some people say] the wetlands are too valuable to be doled out in bits and pieces for agriculture or real estate development." Already, resort development at Fairmont Hot Springs and Windermere Lake has changed the river's course and filled some of its flood plain. The Columbia wetlands are so rich and pristine—and unique in a province whose hydroelectric projects have already flooded many river valleys and their wetlands—that some people want to keep them intact. They believe that nature should be allowed to have the upper hand in the wetlands. The river should be allowed to flood naturally every summer, nourishing the wetlands as it has for longer than any human can remember.

Meanwhile, the British Columbia Ministry of the Environment is promoting a plan to protect and manage the wetlands as wildlife habitat. Under the plan, the government would review all proposals for developing the wetlands to determine potential impact on wildlife. The plan's proponents acknowledge the ecological importance of the wetlands in their natural state. But they believe that, with a little help from humans, the area could support even more waterfowl and other animals than it already does. The plan calls for a range of management techniques, from the simple building of nesting boxes for ducks to the more complicated controlling of water levels in parts of the wetlands. In addition, the plan promotes opening the wetlands to allow more outdoor recreation activities.

Placing nesting boxes in a wetland to increase the presence of waterfowl.

Entering the Wetland Debate

Faced with these competing pressures, the British Columbia Ministry of the Environment and the British Columbia Cabinet join the ranks of government bodies across Canada and the United States. All are asking themselves the same question: What should we do with the wetlands?

Evidence supporting the value of wetlands—both to the integrity of the environment and to our nations' economies—is mounting. So, too, is evidence of the increasingly alarming rate at which wetlands have been drained, filled, flooded and otherwise converted to non-wetland areas. This emerging knowledge sharply focuses the wetland issue on a central concern: How should we balance our need to use and develop wetlands with our need to protect them and the natural benefits they offer us?

In both the U.S. and Canada, wetland policies are being debated, adopted and altered at all levels of government. And the breadth of concerns voiced in the growing public discussion indicates just how difficult it will be to decide what to do with the wetlands. In the discussion that follows, we will explore the wetland debate and provide some options for resolving conflict. After having simmered for a while in both countries, public concern about the issue of wetlands is bubbling more rapidly now in Canada, and is boiling over in the United States. What's behind it?

Important Functions of Wetlands

The World Conservation Strategy, a 1980 document representing the collaborative work of some 450 governmental and nongovernmental organizations, has identified wetlands as key life support systems, which help sustain this planet. What makes them so important?

Flood Control/Groundwater Recharge

Because of their ability to retain water, wetlands can buffer coastal areas from the full effect of storms. They can also slow the progress of flood waters and replenish underground water supplies.

Wildlife and Fisheries

Wetlands support a wide range of plant and animal species, including many rare and endangered species. About 90% of the fish and shellfish harvested in the coastal United States spawn in wetlands. As breeding and nesting areas, and as migratory stopping points, wetlands are important to waterfowl and other birds. Many fur-bearing and game species, such as rabbit, deer, beaver and muskrat, depend on wetlands for food and shelter.

Water Quality

Wetlands take water on a slow journey through an environment rich in plant life and microorganisms. This process helps maintain and improve water quality. Wetland systems can assimilate excess nutrients such as nitrogen and phosphorus from agricultural runoff, allow sediment to settle out of water and process some chemical and organic wastes.

Climatic Stability

Wetlands modify the temperature and moisture content of the lower atmosphere. In some places, they are important contributors to local cloud formation and precipitation. More broadly, wetlands' role in the cycling of gases, such as carbon, nitrogen, oxygen and methane, make them important to the world's atmospheric stability.

Education and Research

Wetlands are living laboratories for a range of scientific research and environmental education programs. These activities may focus narrowly on specific plant or animal species, or on particular wetland functions. Or they may take a broader approach—for example, looking at the connection between wetlands and changes in global climate. Wetlands also provide a place where new species, or new uses for known species, might be discovered.

Recreation and Aesthetics

Wetlands provide recreational opportunities such as hunting, fishing, photography, canoeing and wildlife viewing. They help fulfill people's needs for open space, environmental variety, natural beauty and solitude.

Direct Economic Benefits

Significant income can be generated from the use and harvest of wetland resources. Sources include tourism, recreation, commercial fishing, forestry, peat extraction, fur harvest, wild rice farming, seasonal hay cropping and grazing.

Natural Wetland Benefits

The debate over wetlands has come about in part because we have developed a better understanding of the role that wetlands play in our lives and our environment. We know that wetlands provide breeding and wintering grounds for waterfowl and shorebirds. They offer indispensable habitat for wildlife, and nursery grounds that are the backbone of our commercial fisheries. Wetlands lessen flood damage, recharge the underground sources of our drinking water, and help to control pollution. Their rich and varied landscapes offer recreational opportunities and aesthetic enjoyment.

Putting a monetary value on the many "services," or functions, of wetlands is difficult. But in Canada, government estimates of $5-10 billion in annual benefits make the economic value of wetlands alone a strong argument for using them wisely. These figures include money generated from wetlands-related activities, such as hunting, fishing, trapping, tourism and recreation, as well as estimates of the value of natural functions like flood control and water purification.

The U.S. Fish and Wildlife Service estimates that, in 1985, 141 million U.S. citizens spent $55 billion on wildlife-associated recreation. That figure was up 41% from 1980 totals. Many of these activities, such as hunting, fishing and bird-watching, depend on healthy wetlands.

It is arguable that some of the functions of wetlands cannot, and perhaps should not, be given a dollar value. For instance, wetlands act as one source of moisture for cloud formation, thereby influencing local and regional weather

The first Duck Stamp issued by the U.S. Fish and Wildlife Service.

diversity is critical for life on earth to evolve and adapt to changing conditions. So wetlands—as the home for many species, including species we know to be rare or endangered—may be critical to the long-term health of the environment that sustains us. On a more tangible level, this pool of species—each one the product of thousands or millions of years of evolution—is a resource for our future. From that pool, we may draw new food crops, medicines and other products.

patterns. Rainfall regulated by wetlands can be a key source of water for prairie farmland. And wetlands play a role in *global* climate control: They help to keep atmospheric gases properly balanced by storing carbon, releasing methane and neutralizing nitrous oxide.

In addition, wetlands are enormously productive ecosystems: A shallow marsh can be up to eight times more efficient than a wheatfield at turning the sun's energy into plant matter. Wetlands sustain a wide variety of plant and animal life. Many scientists believe that this biological

The Land Beneath the Water

The value of natural wetlands as related above still does not tell the whole story of what wetlands are worth. Beneath the water that makes wetlands "wet" is land. And that land has value in its own right.

In some cases, that value is contained in the earth itself. Peat, a soil enhancer well known to gardeners, comes from wetlands. Researchers are exploring the potential of peat as a source of fuel, medicine and chemicals. In 1991, the value of peat products from Canada exceeded CAN$100 million. In addition to providing a source of peat, wetlands sometimes lie over minable deposits of other resources such as iron, coal and phosphates.

A marsh (left) can be up to eight times more efficient than a wheatfield (right) at turning the sun's energy into plant matter.

Sometimes the worth of wetlands is affected by their location. Wetland areas in cities, for example, may be the only remaining parcels of land large enough for development projects such as office buildings, storage facilities or recreational fields. And in suburbs, wetlands often offer excellent locations for housing developments or shopping centers.

By far, the most common value attached to the "land" in wetlands is its usefulness for agriculture. Environmental reporter Michael Keating suggests that "some marshes make excellent farms when drained, the rich soil thus exposed being ideal for such crops as carrots and onions. The Holland Marsh north of Toronto was drained in the 1930s and is now one of the richest market gardens in Canada." Not all wetlands make good farmlands. But, from the mid-1950s to the mid-1970s, over 11 million acres (4.4 million hectares) of wetlands were converted to agricultural use in the United States alone. And over 85% of wetlands lost in Canada since European settlement in the early 1800s is due to agricultural drainage.

Some wetlands are altered for reasons other than the value of the dry land beneath them. For example, flood-control levees built along the Mississippi River have cut off delta wetlands from the river. Without the sediment that the river naturally deposited in these coastal wetlands, they are becoming submerged and turning into open-water areas. Other wetlands dry out because their sources of water are decreased. Large reservoirs and water diversions on rivers in the western United States, for example, can cut off water to wetlands farther downstream.

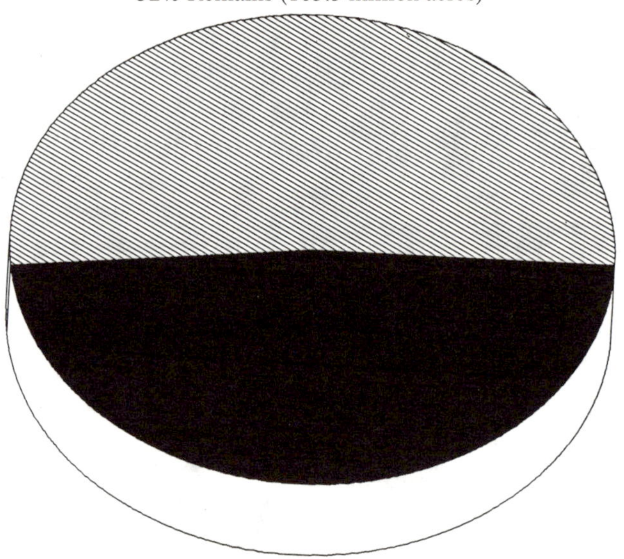

The U.S. Fish and Wildlife Service estimated in a 1983 report that over 200 million acres of wetlands existed in the lower 48 states at the time of European settlement.

A channel was dug to drain this wooded wetland area near Virginia Beach, Virginia.

Cumulative Effects

It is clear that both the United States and Canada benefit from converting wetlands to new uses. In places where wetlands once stood, there are now farmlands that exist where cultivation was impossible before, residential developments that provide needed homes for a growing population, new highways and airports that make travel easier, expanded ports and canals that enhance commerce and levees that help keep our rivers on reliable courses.

But there's the rub. The wetlands that have been so altered *are no longer wetlands*. In gaining benefits of conversion, we lose the benefits of natural wetlands.

As both the United States and Canada were developed, most of the beneficial functions of natural wetlands were not recognized. These benefits didn't get factored into the equation when it came time to figure out what to do with the wetlands. Converting wetlands to other uses seemed more productive than keeping them in their natural state.

Government policies often promoted the view that wetlands were unproductive and undesirable in their natural state. The Swamp Lands Act, for example, passed in 1850 by the U.S. government, was designed to bring more land into cultivation. Under this law, the U.S. government turned over 70 million acres (29 million hectares) of "swamp and overflowed lands" to the states. In the eyes of the federal government, it was a trade. The states would take ownership of large acreages within their boundaries. All they had to do in return was to make the land suitable for "productive" use.

In the development of our countries, wetlands were increasingly filled and drained and converted to other uses—primarily agriculture. It is only now that

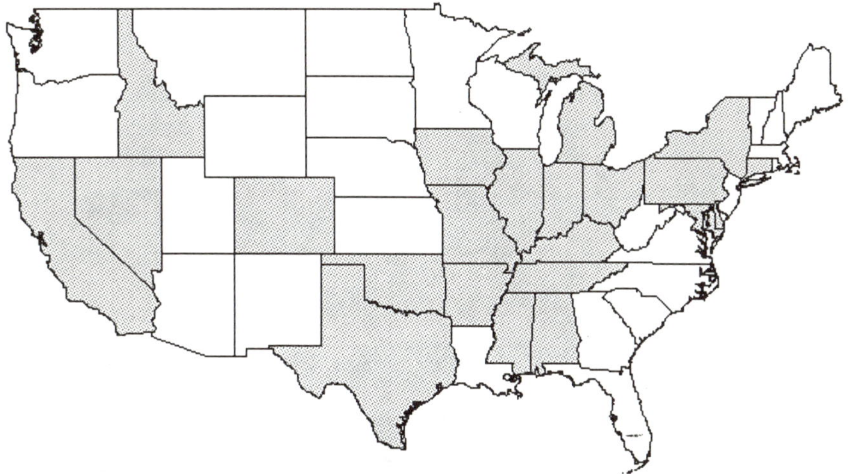

States that lost more than 50% of their wetlands between the 1780s and mid-1980s

State	Percent Loss
Alabama	50
Arkansas	72
California	91
Colorado	50
Connecticut	74
Delaware	54
Idaho	56
Illinois	85
Indiana	87
Iowa	89
Kentucky	81
Maryland	73
Michigan	50
Mississippi	59
Missouri	87
Nevada	52
New York	60
Ohio	90
Oklahoma	67
Pennsylvania	56
Tennessee	59
Texas	52

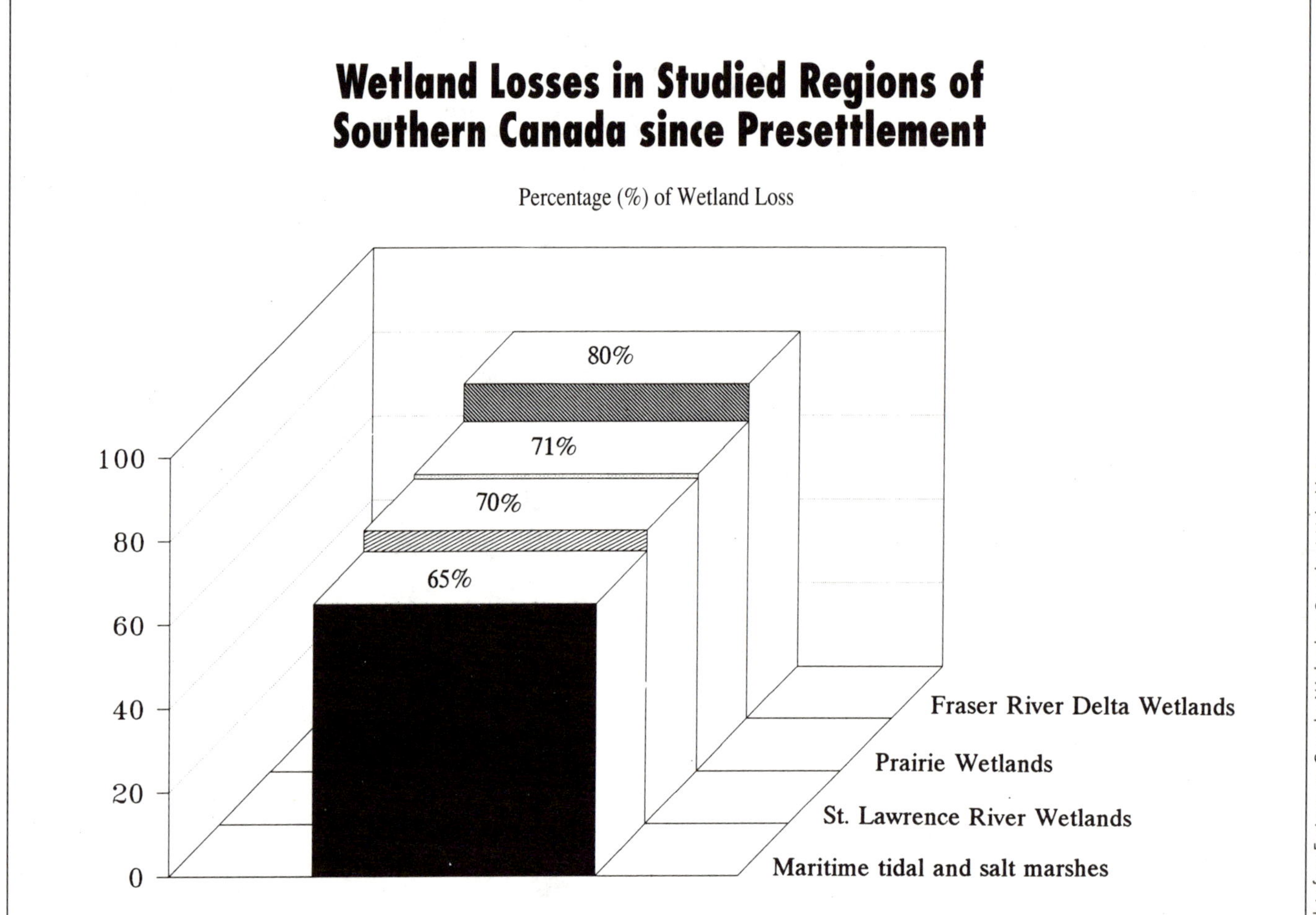

Wetland Losses in Studied Regions of Southern Canada since Presettlement

the consequences of this wetland conversion are being calculated.

Recent studies estimate that about 48% of the wetlands in the lower 48 United States and around 14% of Canadian wetlands that existed when settlement of our countries began are no longer wetlands. In 1992, wetlands covered about 314 million acres (127 million hectares) in Canada. That is, 14% of its total land area. About 5% of U.S. territory is covered by wetlands, an area of about 103 million acres (42 million hectares).

The wetland story isn't quite as clearcut as these figures about loss and remaining wetlands suggest. They do not reflect the fact that many of the areas that remain have been polluted and misused. This kind of abuse reduces a wetland's ability to function fully.

[Note: Figures on wetlands and losses in the United States usually include only the lower 48 states. This book follows that practice. Adding in the vast and mostly pristine wetland areas of Alaska— some 200 million acres (81 million hectares)—would skew the picture for the rest of the country. In addition, the wetlands of Alaska and northern Canada are often considered separately in policy debates. They are largely undeveloped, and, since the populations near them tend to be small, there is less pressure to develop them. The Canadian wetland debate often focuses on the southern, most populated parts of Canada. This is where wetland conversions—and conflicts over their use—have been most significant.]

Today's Challenge

As we have learned more about the value of natural wetlands, we have begun to see our decisions about their use in a different light. When we knew less about the benefits that natural wetlands provide, the tradeoffs made in converting wetlands to other uses seemed much less significant than they do today. What should we do with wetlands? The answer is no longer as clear as it was in the days of the Swamp Lands Act. The seriousness of these tradeoffs prompts us to take a hard look at our present practices.

Conversion of wetlands continues at alarming rates in both countries. The wetlands that disappeared between the mid-1950s and the mid-1970s in the United States would cover an area three times the size of New Jersey. That means that over 15 million acres (6 million hectares) of wetlands were converted to some other use during that period. And studies estimate that between 275,000 and 500,000 acres (up to 202,350 hectares) of wetlands continue to be altered each year in the United States. When one considers that only about 25,000 acres (10,120 hectares) are created or restored each year, these conversions result in a sizable yearly dent in our wetland resources.

At the same time, popular and official support for protecting wetlands is growing. Especially over the past decade, international agreements, national policies, provincial and state initiatives, and private efforts have laid a foundation of recognition for the values of wetlands.

In 1986, Canada and the United States signed the North American Waterfowl Management Plan. Later, Mexico became involved. Under the Plan and related agreements, the three nations agreed to protect and improve millions of acres of wetland habitat. These areas are critical to the survival of the continent's migrating ducks, geese, and swans. All three countries have also signed the Ramsar Convention on Wetlands of International Importance. Under this accord, they join over 70 other nations in identifying and protecting globally significant wetlands.

On a more local level, a long list of swamps, marshes, and bogs—from Priddis Slough near Calgary to Delhomme Marsh on the Gulf of Mexico to Manitoba's Oak Hammock Marsh—have attracted public attention for protection. Private organizations such as Ducks Unlimited, The Nature Conservancy, and a variety of local groups have been active in acquiring and preserving wetlands.

The wetland challenge that faces us today is to decide what kinds of tradeoffs are acceptable. How should we weigh the benefits of wetlands against the benefits of their conversion? What can we afford to give up? Are current efforts to protect wetlands sufficient? Who should make these decisions?

Primary Causes of Wetland Loss and Degradation

Human Impacts
- Drainage
- Dredging and stream channelization
- Deposition of fill material
- Diking and damming
- Tilling for crop production
- Grazing by domesticated animals
- Discharge of pollutants
- Mining
- Alteration of hydrology (water flow)

Natural Threats
- Erosion
- Subsidence
- Sea-level rise
- Droughts
- Hurricanes and other storms
- Overgrazing by wildlife

Source: U.S. Environmental Protection Agency

Creating and Restoring Wetlands

Mitigation is a strategy used in many government wetland policies and proposals. It is a way of making up for the loss of wetlands—or their functions—that takes place when wetlands are converted to other uses. A permit to fill a wetland, for example, may require the creation of a comparable wetland, usually in a nearby location.

The three methods used to compensate for both current and former wetland losses are enhancement, restoration and creation of wetlands. Enhancement is improving certain conditions in existing wetlands. An example of enhancement is building levees to control water levels in a marsh, which may enhance an area's ability to support ducks. In other cases of mitigation, restoration may be required to repair damaged wetlands. Reestablishing water flow to a drained area is one example of restoration. And creating a wetland where none existed before results in the most drastic landscape changes of the three methods.

One of the main questions about mitigation as a workable part of our policy is, "How reliably can we replace the natural wetlands we alter?" Wetland mitigation is a young science, which has gained popular interest only in the last decade or so. In 1989, the U.S. Environmental Protection Agency, in conjunction with the Association of State Wetland Managers, issued a report entitled *Wetland Creation and Restoration: The Status of the Science*. The report sheds light on how much scientists understand and on the success of restoration and creation projects.

Here is an overview of some of its findings:

1) We have a limited amount of experience with creating and restoring most types of wetlands. Scientific literature is similarly limited. What is available varies with the type of wetland and with the region of the country.

2) Creating or restoring a wetland that exactly replicates a naturally occurring wetland is impossible. Natural and artificial wetlands are complex and varied. The intricate relationships among all parts of the ecosystem cannot be duplicated. However, it is possible to come close for some types of wetlands. And certain wetland functions can, in some cases, be created.

3) For a wide range of reasons, partial failures are common for creation and restoration projects. Total failures have been common for certain types of wetlands, including seagrass beds and forested wetlands. Projects on coastal, estuarine and freshwater marshes have been more successful.

4) The ability to restore or create the conditions that yield particular wetland functions varies. Some functions, such as providing waterfowl habitat, are well researched and relatively easy to create. Others, such as the ability to replenish underground water supplies, are much more difficult to assess and create.

5) Long-term success may be quite different from short-term success. Success is often measured by looking at the area's plant life. But revegetation of a restored or created wetland over a short period of time is no guarantee that the area will continue to function over the long term. Natural wetlands may be better able to withstand changing conditions than artificial wetlands.

6) Restoration should be favored over creation. Restored wetlands usually are more successful than created wetlands at recapturing the full range of wetland functions and at surviving over the long term.

The newness of the science of wetland mitigation means that there are significant gaps in knowledge, experience and opportunities for long-term observation. Even people with a lot of experience in wetland restoration, like Dr. Ed Garbisch of the nonprofit restoration company, Environmental Concerns, Inc., acknowledge that wetland creation and restoration is tricky business. But, with 350 restorations to his credit, Garbisch maintains that success is possible. With a growing demand for this kind of work, he says, "The [biggest] problem is that there are too many people getting their hands on it who shouldn't be in it."

Sources: *Wetland Creation and Restoration: The Status of the Science, Vol. 1.* October 1989, U.S. EPA, summary by Jon A. Kusler and Mary E. Kentula; and "Restoring Lost Wetland: It's Possible But Not Easy," in *The Washington Post*, by William K. Stevens.

The Need for Coherence

Canadian and U.S. governments and citizens are turning a critical eye toward government policy about wetlands. What they find in both countries—and in many states and provinces—is a hodge-podge of laws, programs, regulations, and guidelines that somehow affect the use of wetlands. These policies have been instituted at different times, and for different purposes. Taken together, they do not send the same message of concern about the loss of wetlands that more and more citizens are voicing. Nor do they amount to a coherent approach to making decisions about wetlands. Many of these laws and regulations are criticized for being inefficient, contradictory, and unfair to landowners, and for lacking a common goal.

A report published by the Environment Council of Alberta in 1990 concludes that "there are eight federal and 15 provincial Acts or policies that either directly or indirectly affect the management of Alberta wetlands. Much of this legislation is contradictory and has resulted in a complicated web of competing policies that is confusing for regional and provincial agencies as well as for the public." The statement reflects what is happening in the United States, as well. A brief look at some of the U.S. and Canadian laws and regulations that affect wetlands demonstrates the range of approaches and programs that are now in use.

In the United States, several pieces of federal legislation have direct influence on wetland decisions. Chief among these is section 404 of the U.S. Clean Water Act, which requires landowners and developers to obtain government permits for dredging or filling in wetlands.

Weighing benefits of wetlands vs. benefits of converting them to other uses.

Section 404 regulates only about 20% of the activities that alter wetlands. Its reach does not extend to many causes of wetland conversion and degradation, such as draining, flooding, and pollution. Also, certain activities are exempt from the law, including most day-to-day farming, ranching, and forestry practices.

The 1985 and 1990 Farm Bills established two major wetlands programs that work differently from section 404. The "Swampbuster" provision tries to discourage farmers from converting wetlands to agricultural use. It permanently denies federal farm program benefits to anyone who does so. And the Conservation Reserve Program (CRP) encourages farmers to remove highly erodible land from production in return for yearly payments from the government. As of 1991, an estimated 410,000 acres (170,124 hectares) of wetlands were enrolled in the CRP program. The annual cost to the government is about $20 million dollars.

Other legislation, such as the Water Bank Act and the Migratory Bird Hunting and Conservation Stamp Act, has created programs aimed at wetland conservation.

In March of 1992, the Canadian government adopted the Federal Policy on

Bob Hora

Wetland Conservation. It promotes national efforts to protect wetlands and guides the management of federal lands. Canada has not set up regulatory programs similar to section 404 of the U.S. Clean Water Act. But in other respects, its range of federal laws that affect wetlands is similar. These laws include the Migratory Birds Convention Act, the Canada Wildlife Act, the Federal Water Policy, and the Federal Policy on Land Use.

These programs are all aimed at protecting wetlands. But some policies encourage the conversion of wetlands.

The Canadian Income Tax Act, for example, provides tax deductions for expenses incurred while draining wetlands. And some Canadian government agricultural programs offer income and marketing supports based on the amount of land a farmer cultivates. These programs encourage farmers to cultivate more of their land to get more government aid. So even marginal land, often wetlands, is put under the plow.

Many government programs can work either way—encouraging the protection *or* the conversion of wetlands—depending on how they are implemented. A look at the Alberta Planning Act illustrates this point. The act allows the provincial cabinet to set up "special planning areas" so that resources of public importance can be controlled. This option could be used to protect wetlands, but usually it is not. Also, under this same act, municipal governments are given the authority to protect wetland areas within proposed subdivisions. But these local governments also have a conflicting authority: They are able to approve and fund the draining of wetlands.

Definitional Debates

Any government program designed to regulate the development of wetlands must answer the question: What areas will fall under the government's control? Agreeing upon a definition for these "jurisdictional wetlands" is no easy task. Officials in the U.S. White House and several federal agencies can attest to that.

Figuring out what is and what is not a wetland is sometimes a tough job. Wetlands come in an enormous variety of types. Their water level may change over the course of the year. They may undergo other changes over a longer time. They occupy a "transitional" zone between open water and dry land environments. All of these characteristics often make it hard to determine their precise boundaries.

Deciding which wetlands should come under government jurisdiction adds another level of complexity. This decision is central to the U.S. debate over wetland policy and is sharpened by the fact that there are four federal agencies responsible for identifying wetlands under existing regulatory programs. Until recently, these agencies used different guidelines and procedures.

In 1989, the Army Corps of Engineers, the Environmental Protection Agency, the Fish and Wildlife Service and the Soil Conservation Service adopted unified guidelines for making wetland determinations. This new *Federal Manual for Identifying and Delineating Jurisdictional Wetlands* reflected agreement on three essential criteria: plant life adapted to water-logged, low-oxygen soil conditions (hydrophytic vegetation); soils whose chemical make-up has been altered by the presence of water (hydric soils); and certain patterns of flooding or saturation (hydrology).

Almost immediately, the guidelines came under fire on some fronts for casting the government's jurisdictional net too broadly. Critics complained that the guidelines included areas that did not appear to be wetlands, or that did not seem to function like wetlands. The next two years were spent trying to clear up confusion over the manual and field testing it. And the White House Domestic Policy Council got involved by gathering public comments on wetland policy and regulations from across the country.

Revisions to the manual were proposed in 1991. These proposals have drawn fire from another direction. This time critics are concerned that the revised guidelines would drastically reduce the amount of wetlands under government jurisdiction. They charge that the changes would allow development of ecologically important areas. While the proposed changes include a variety of modifications, public controversy seems to center on two factors: the length of time that land must be saturated or covered with water to qualify as a wetland, and how close the water must be to the surface in order to indicate saturation.

The debate over these guidelines illustrates the difficulty in defining jurisdictional wetlands. The challenge lies in the fact that there are no universally accepted or scientifically precise answers to the questions of where wetlands begin and end, and which wetlands are significant enough to regulate. Science can only give guidance to the political process that must eventually decide on the proper reach of government control.

CHOICE #1

. . . rests on the belief that we should focus our policy decisions on the total array of benefits wetlands provide us—and on creating the greatest benefit for society. Choice One supporters also believe that, although we do have some data on existing or altered wetlands, our knowledge is far from complete. Therefore, we need a thorough inventory of our wetland resources in order to make informed decisions. These inventories will help us understand where our wetlands are. And they will point out the natural functions of wetlands—such as water purification, wildlife habitat and recreational opportunities. Armed with this information, we will be better able to weigh the benefits of preserving specific wetlands against the benefits of altering them or managing them to enhance certain functions. This perspective argues that government must take the leading role in obtaining public input and making these decisions.

CHOICE #2

. . . relies on landowners to make wise and responsible decisions about wetlands. Advocates point out that most wetlands in the continental United States and southern Canada are privately owned. They believe that government should stay out of the business of deciding what landowners can and cannot do with their wetlands. Instead, government should work with these landowners to promote private stewardship of wetlands. Proponents of Choice Two believe that economic incentives, such as tax deductions or direct payments for wetland protection, should anchor government programs. They believe that, especially if there is a strong program of public and landowner education in place, the private sector will make informed decisions about land use. Supporters of this approach believe it offers an effective conservation strategy that preserves our countries' economic vitality by protecting the rights of property owners.

CHOICE #3

. . . maintains that we should permanently preserve most remaining wetlands. Supporters of this choice point to the enormous destruction of wetlands and to what we are learning about their natural value. They argue for a strong protection program based on government purchase and preservation of wetlands, strategies that persuade private landowners to permanently protect their wetlands and broad government regulations. They also support efforts to restore altered wetlands to their natural state to heal some of the damage we have already inflicted. Advocates of this choice argue that a policy of cooperation with nature—rather than domination over it—is the most prudent and spiritually satisfying course. They believe that public education programs coupled with opportunities to visit and enjoy wetlands will spark the insight that we would need to change our human-centered approach to wetland decisions.

A Public Challenge

It is widely agreed that our often-contradictory wetland policies and programs need a guiding force. They need a common direction to help uncover and correct contradictions, to govern the use of existing programs and to point the way toward creating effective new programs. Beyond that basic agreement, however, there is little consensus about how to strike the right balance in wetland policy. Especially in heated conflicts, it often seems as though the debate boils down to two positions: preservation and development.

But sifting through the arguments and proposals on the table shows that the issue is much more complex than that. In the discussion that follows, we will look at three different and compelling perspectives on wetland policy. Each perspective (we call them "choices") holds a unique view of the problem. Each proposes a distinctive approach to its solution.

MAKING WETLANDS WORK FOR US

In 1986, the northern California community of Arcata became the first U.S. city to build a sewage treatment plant that looks suspiciously like a marsh.

For a long time, the city had depended on oxidation ponds—a traditional sewage treatment method—to clean up its waste water enough to allow the water to be dumped into nearby Humboldt Bay. But when California upgraded its sewage-discharge rules, Arcata discovered that these ponds did not have the reliable cleaning power needed to meet the new standards. And the state had clear ideas about how Arcata should treat its sewage in the future: It was planning a new regional sewage plant for the area and wanted Arcata to tie in to help foot the $50-million construction bill.

But Arcata came up with a different approach. After partial treatment in the oxidation ponds, the water takes a two-month journey through 154 acres (371 hectares) of artificial wetlands. In the marsh, the water comes in contact with hungry microorganisms that live around the roots and stems of the marsh plants and help purify the water. After this encounter, it meets the state's discharge standards. It is also generally cleaner than the water already in Humboldt Bay.

Arcata created an effective solution to its sewage treatment problems and paid less than it would have for higher-tech approaches. At the same time, it turned an abandoned area of waterfront into a vibrant—and pleasant-smelling—marshland park. Its trails, benches, and bird blinds bring Arcatans into direct contact with the natural world of the wetlands, and with the stunning variety of wildlife and plants they attract and support.

Arcata's creative use of wetlands focuses our attention precisely where advocates of our first choice believe it should be—on the benefits that wetlands can provide. According to this view, we should avoid the emotional heat that often surrounds environmental issues. We should think with clear heads about how to use wetlands in the best interests of society.

The Arcata Marsh, with its associated primary wastewater treatment plant, oxidation pond and wildlife sanctuary.

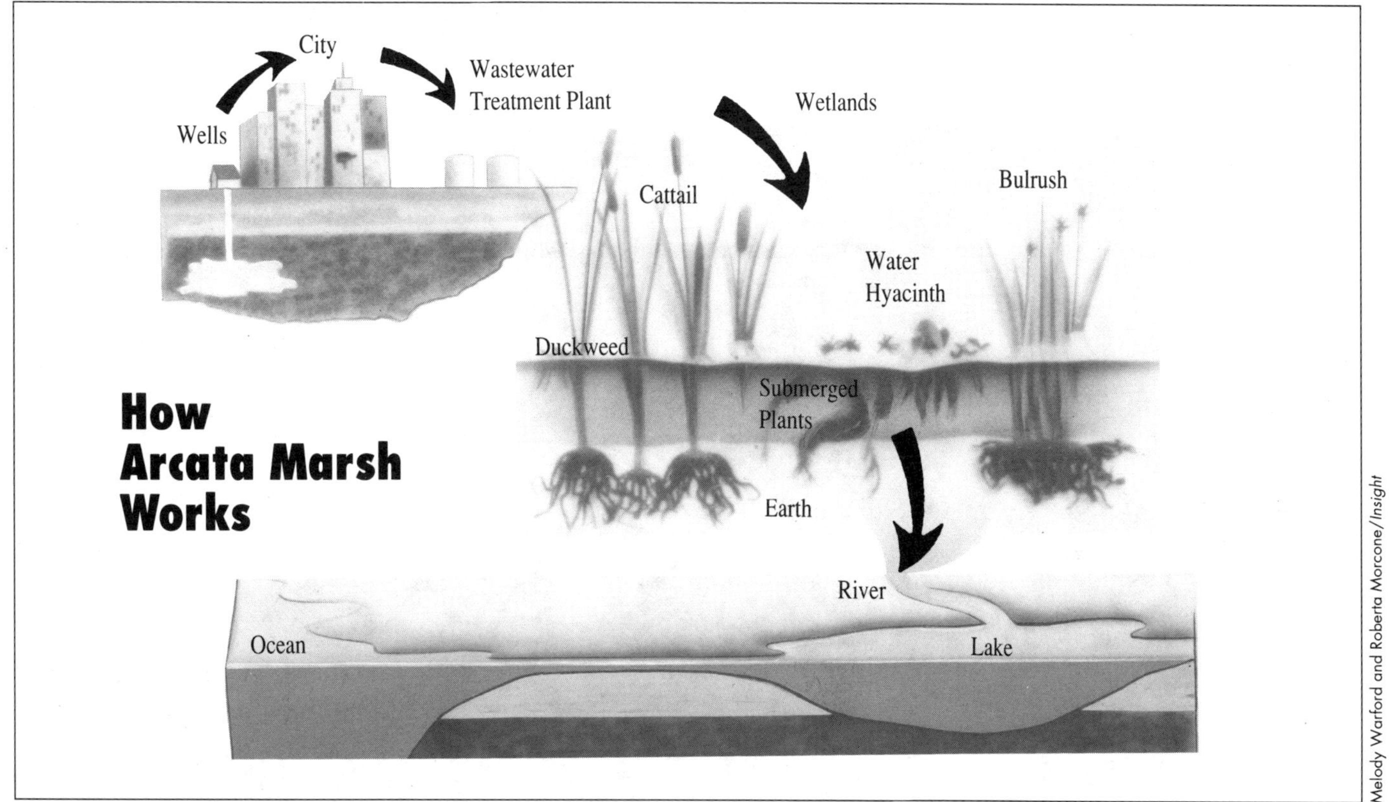

How Arcata Marsh Works

Useful Resources

Former U.S. President George Bush used a simple phrase to capture his new policy toward wetlands: "No net loss." But people like Ronald Anderson, president of the Louisiana Farm Bureau, believe that "'No net loss' was a catchy campaign phrase that represented good politics, but bad policy." Supporters of Choice One are concerned that, under this goal, our efforts will be directed at every area that fits the scientific definition of a wetland.

Advocates of this choice say that this scientific definition is not enough to tell us what to do with wetlands. For them, knowing that an area is wet for a certain length of time each year or knowing that the plant life is adapted to wet conditions goes only part of the way toward defining a wetland. They believe that we need to know what each wetland does to know what we should protect. Does it provide wildlife habitat? Does it purify water? Is it beautiful to look at? These are some of the functions of wetlands. Some wetlands are better than others at providing certain functions. So people who agree with

Choice One argue that we have to focus on what each wetland does or we could end up spending a lot of money to preserve areas that provide little benefit to society.

Supporters of Choice One say that "conservation" should be the watchword of wetland policy. They define conservation as "the wise use of resources." In their view, wetlands are resources that we should use for our benefit, but also protect for future generations to use. This perspective advocates a variety of approaches for managing our wetlands. The approach we use should depend on the kind of wetland we are managing and on what is most important to society about that wetland.

Advocates of this option believe that certain wetlands should be preserved in their natural state. Some wetlands, for example, may be especially fragile; they may be good examples of a certain kind of wetland; or they may provide a very important function, such as flood control near a city. Other wetlands do not need as much protection. They can support a variety of human uses, from seasonal hay cropping to commercial trapping to recreation, without being damaged.

In many cases, supporters of Choice One agree with Nelson Wilson, president of Land Improvement Contractors of America (LICA). He believes that "[i]t makes no sense at all to identify an environmental value that a wetland provides and not try to enhance that value [and use the resource to our best benefit]." According to this perspective, managing or altering wetlands can sometimes enhance the benefits they provide. Proponents of this choice point to numerous wildlife refuges throughout the United States and Canada that actively manage wetlands—by means such as building dikes and controlling water flow—to make them into even better habitat for waterfowl.

People who agree with this first choice believe that some wetlands are of little value in their natural state. These areas may actually offer more benefit to society if they are drained for agricultural use, filled for housing developments and new stores and office buildings or mined for peat and other natural deposits. And sometimes, as the Arcata story suggests, wetlands can be created or restored to serve valuable functions.

Peat: A Fuel for Our Future?

A peat farmer in Canada.

Bogs. These strange wetlands have always haunted us. Misty moonscapes of mosses that quake underfoot . . . . Bogs can be found throughout the world, and everywhere they have been regarded with suspicion and dread.

Louise E. Levathes

Bogs may have long been a source of fear, but they also have long been a source of fuel. A time-honored fuel for rural homes around the world, peat is now being eyed for its potential to create energy on a much larger scale.

Peat is a thick layer of partly decomposed plant matter that has accumulated over thousands of years. Over an immense amount of time, peat would eventually turn into coal. But we do not have to wait that long in order to use it as an energy source.

An article in *Canadian Geographic* estimates that Canada's northern peat bogs hold the energy equivalent of four billion tonnes (4.4 billion tons) of oil. As other sources of fuel, such as oil and natural gas, are used up or become more expensive, peat is becoming a more attractive option.

Several European countries already use peat to generate electricity. And in Maine, the first U.S. peat-fueled power plant has supplied electricity to Boston's suburbs since 1989. Peat for this power plant is harvested from the Down East Peat company's Denbo Heath. This thousand-acre bog is part of Maine's estimated 750,000 acres (303,634 hectares) of peatlands. Frank Graham writes in *Audubon* magazine that "large machines dubbed 'peat-poopers' cruise the bog, pulling the top couple of inches of peat onto a maceration drum, which feeds it back by conveyors to screw augers. The sods are then extruded through cylinders to the surface of the bog, where they lie in long sections, like lengths of hose, to dry in whatever sun the Maine skies are able to produce."

Using peat to produce electricity could become big business in some parts of the United States. In addition to Maine's peat reserves, Minnesota has about seven million acres (2,832,861 hectares). And other states, including Florida, Michigan, Illinois, Colorado and Indiana, have peat as well.

Even without its contribution to energy production, peat is already a $74 million industry in Canada. Harvested in nearly every province, peat is stripped from the bog, dried, and sold as a soil conditioner.

In weighing the promise of peat as a large-scale energy source, significant tradeoffs also must be taken into account. Critics argue that harvesting peat at levels needed to support this production would be disastrous. They point out that, in many European nations, high demands for peat to fuel energy plants have decimated peat deposits. A similar situation could occur in the United States and Canada, where current harvest levels are sustainable over the long term.

Sources: *National Geographic*, *Canadian Geographic*, and *Audubon*.

Taking Stock

Advocates of our first choice think that the government should conduct an inventory of existing wetlands. This inventory would help us identify wetland sites with very high or special environmental value. It would point out areas that are in danger from pollution or misuse. And it would identify which types of wetlands are rare or decreasing in number within a region.

This wetland inventory should help government decide how to manage each wetland area. Proponents of Choice One suggest three basic categories of government control:

- **Preserve wetlands of high environmental value**. Government purchase of special wetlands and incentives for private landowners to preserve their wetlands.

- **Allow unrestricted use of wetlands of low value in their natural state**. Landowners would be able to use these areas in the same way as they would use any other land.

- **Ensure that all other wetlands are used in a way that balances different needs of society**. Government should set up regulations to make sure that decisions about wetland use are made with input from the public. These decisions should consider a range of factors, including the environmental characteristics of the wetland, and social and economic needs.

. People who take this perspective believe that, with a plan like this in place, we would direct the proper amount of attention and resources toward each wetland decision.

Scientist analyzing wetland soils.

Public Choices, Government Roles

With the research and the general policy framework in place, proponents of this view feel that making decisions about wetland use becomes a political question. Advocates of this option believe that these decisions affect our whole society. So they think that there should be ample opportunity for public input into them. In their view, scientific reports and government panels cannot speak for the public. To get public input, the government must carry out an on-going program of hearings, comment periods, and informational meetings. Their aim must be to involve people in setting priorities for the use of wetlands.

According to this perspective, local and state or provincial governments are in the best position to carry out much of this political work. Advocates of this choice believe that local or regional decision-making can best focus on the unique conditions of each situation. And, supporters say, with their direct contact with the citizenry, local levels of government can more easily promote active citizen participation.

In most cases, proponents of Choice One believe that the federal government should set flexible guidelines for wetland conservation. Because these guidelines can be adapted to different local situations, they will work better than strict national regulations. Also, by providing technical assistance to states, provinces and localities, the federal government can help create an effective, manageable and flexible program for conserving wetlands that meets local needs.

Making Up for Losses

Proponents of this choice assert that, in many cases, development that alters or destroys wetlands is, nonetheless, desirable. A proposal for a new marina or a highway may require filling in some marshes. The best place for a new shopping center or apartment building may be on a section of swampy land. But sometimes this kind of development will damage wetland functions that are economically, socially or ecologically important to protect, such as wildlife habitat or flood control. So Choice One supporters rely on a system called *mitigation*, or compensation, to balance those losses.

Mitigation compensates for lost functions of wetlands by combining four strategies: Creating new wetlands, restoring areas that were once wetlands, enhancing existing wetlands to improve certain functions, and permanently preserving unprotected wetlands. With the proper legislation in place, governments can require mitigation as a condition before permits to develop in wetlands are granted. Florida, for example, began reviewing proposals for wetland development in 1979. The permits it issues often require some kind of mitigation to offset wetland losses. Ann Redmond of the Florida Department of Environmental Regulation (DER) notes, "From January 1, 1985 through December 6, 1990, the DER issued 1,262 permits that included requirements for the creation, enhancement, or preservation of wetlands. The permits authorized the loss of 3,305 acres of wetlands, and required that 3,345 acres be created, 7,301 acres be enhanced, and 7,588 acres be preserved." Proponents of this choice believe that mitigation gives the government a powerful tool for protecting the benefits that society gains from wetlands.

Enhancing Wetland Benefits

"The conversion of a wetland does not necessarily result in the loss of a wetland. To the contrary, a well designed conversion results in the enhancement of wetland values." With these words, Nelson Wilson challenged a group of U.S. legislators to rethink their ideas about wetland protection. Wilson, president of Land Improvement Contractors of America, talked about ideas that make a lot of sense to supporters of Choice One. His main question was: Why be satisfied with preserving wetlands, when we can actually improve on nature?

"The party line on wetland policy says that wetlands must be protected because they provide: (1) flood and erosion control, (2) aquifer recharge, (3) groundwater filtration, and (4) wildlife habitat. If these are, indeed, the values our wetland policy seeks to protect, why should we stop with protection? Each of these values can be greatly enhanced through the use of conservation technology. Flood and erosion control can be greatly enhanced through the use of terraces and water retention structures. Aquifer recharge is much more effective when surface waters are routed to recharge impoundment areas. Groundwater filtration is up to 30% more effective simply by installing subsurface water table management technology. Wildlife habitat can be transformed from a mosquito infested, snake, rat, and vector breeding ground to a safe, aesthetically pleasing wildlife sanctuary. . . . The current policy of protection by preservation . . . denies the public the environmental values that conservation technology offers.

"My assertions are not idle claims or wishful thinking, as some would have you believe. Since the mid-1950s, the PL-566 Watershed Program has amassed an indisputable record of enhanced environmental values . . . in watersheds across America. Our wetland policy should embrace this technology and encourage its expanded application. Current policy seeks to preserve wetland areas; it should seek to enhance environmental values."

Source: *The LICA News*, November 1991, comments presented by Nelson Wilson to the House Water Resources Subcommittee.

*A **wetland mitigation project** (dark section in center of photo).*

Choice One supporters admit that wetland creation is a new and experimental field. To a lesser extent, so is restoration. But they maintain that innovations in these areas will improve our ability to duplicate, and even to enhance, the things that we find most valuable about wetlands. Advances in our ability to replicate natural wetlands, they believe, is only a matter of time and experience. With the commitment of government and private research funds, they feel scientists will make great strides in increasing the scope of knowledge about how to better enhance and restore wetlands, and create new ones.

For now, proponents suggest, governments should not rely on creation of new wetlands as the main form of mitigation. Instead, we should depend on more proven methods, such as restoration or enhancement. As our knowledge grows, we will be able to rely more on creating wetlands as a form of mitigation.

Knowledge Is the Key

People who agree with this option believe that our decisions about wetlands can be no better than the information on which they are based. As Dr. Bruce D.J. Batt, director of the Institute for Wetland and Waterfowl Research, says, "Our philosophy is quite similar to that found among businesses everywhere—a relatively small initial expenditure for 'R & D' [research and development] can pay off in big dividends later."

The "big dividends" that Choice One supporters are looking for are wise decisions about managing wetlands. Their research program aims to clear up uncertainties and fill gaps in our knowledge of wetlands. They also want to get the research results into the hands of policymakers, wetland managers, landowners, the general public and others involved in making wetland decisions. Government funding for research must be targeted at the following areas:

- Mapping existing wetlands.
- Understanding the condition of existing wetlands and the rate and causes of their destruction. This is particularly vital in areas of rapid loss or high value.
- Understanding the functions of different kinds of wetlands.
- Evaluating current techniques for protecting and managing wetlands.
- Testing and evaluating methods of creating and restoring wetlands.

In the eyes of Choice One proponents, our best approach to wetland policy works with our knowledge and our ability to manage natural systems to produce the maximum benefit for society.

Restoring a wetland area.

U.S. Fish and Wildlife Service

What Opponents Say

People who disagree with this approach argue that it is unrealistic to believe that people can adequately manage wetlands. What's more, they think it is arrogant to assume that we can replace the wetlands we destroy. Critics of Choice One see wetlands as complex and varied natural systems that we will never fully understand nor be able to duplicate. Some opponents argue that because so many of our wetlands have already been destroyed, we should go no further along the same path. Degrading or destroying the natural wetlands that remain, they say, is a high-stakes gamble with nature that we should not take.

Other critics argue that this approach concentrates on what wetlands do for humans. To them, wetlands are more than just the sum of their social benefits: Wetlands play a critical role in the natural balance that sustains all life on earth. From this perspective, making wetland choices subject to political and local decision-making disregards the larger importance of these ecosystems. Canada alone encompasses nearly one-quarter of the world's wetlands. That fact carries with it global responsibilities.

Opponents of this choice also charge that this approach creates a cumbersome and expensive system for making wetland decisions. It examines each decision in too much detail and takes too much time to be effective. In addition, it demands an unrealistic level of public involvement. All this, they say, is done under the guise of providing a way of managing resources efficiently. Some critics also fear that local governments would be unable to make balanced decisions about wetlands. And they are concerned that the money and clout of pro-development interests would influence decisions.

Other opponents object to what they see as a huge government role that could easily get out of hand. They believe that this approach to wetlands policy gives the government free rein to instate a huge and unnecessary land-use regulation program. Such a program, they argue, would slight the rights of property owners in the name of benefiting society.

Our second choice focuses on this concern for the rights of property owners. Choice Two advocates believe that we should rely on the wise decisions of landowners as the basis for our wetland policies. Let's consider their perspective.

CHOICE 2

When Cliff Gardner's ancestors settled in Nevada's Ruby Valley in the 1860s, they found a harsh, dry land overgrown with sagebrush. But they, and other pioneers like them, made that land productive. They built rock dams to catch the river as it flowed into the valley. And they diverted the water across the land in a network of irrigation ditches. Now, over 100 years later, their descendants—ranchers like Cliff Gardner and his family—run livestock and grow crops on the same land.

The agricultural practices that have allowed these Nevada ranchers to make a living for decades have also, some say, enhanced the natural surroundings in a vast area of land by increasing plant health and this western region's range of plant communities. Gesturing across irrigated pasture land, Gardner points out to a visitor the green areas and wetlands that he says were created, and are maintained, by day-to-day ranch operation. These areas support an amazing diversity of plant and animal species more suited to wet areas than to the native, desert-like surroundings.

Proponents of our second choice believe that landowners like Cliff Gardner have been making wise decisions about their land and resources for generations. These owners realize that taking care of their land benefits them—and their ability

Dry land in the western U.S. turned green by the irrigation of settlers and landowners.

Cliff Gardner

"

to profit from the land—over the long term. Advocates of Choice Two point out that most wetlands in the 48 contiguous United States and in the southern, most-populated portion of Canada are privately owned. Successful wetland policies, they believe, will respect the rights and competence of private landowners.

In this view, government interference in the decisions of property owners violates the spirit of individual initiative and enterprise that built our countries. As much as ever, our societal well-being depends on a strong economy. A strong economy, say supporters of Choice One, is based firmly in the freedom to use privately owned property for economic gain.

Ridiculous Regulation

Advocates of Choice Two acknowledge that wetlands are important to society as a whole and that we may need strategies to protect them. But when we create these strategies, they want to avoid the kind of disregard for private property they see currently taking place. In the United States, the federal government has the authority to regulate or even to prevent certain alterations of privately owned wetlands. Proponents of this choice charge that this authority extends to an astonishing array of areas, including those that may not seem like wetlands at all and to tiny tracts of land that appear insignificant.

Richard Adamski, a retired state trooper from Baltimore, ran smack into this regulatory wall when he wanted to build his retirement home in a small Maryland town. He owned a three-quarter-acre forested lot in an already-developed residential area. Unfortunately for him, his

A "Multiple Use" Perspective

A movement that is rapidly finding its feet in the United States and Canada, the "multiple use" or "wise use" movement is an increasingly vocal collection of landowners, property rights advocates and outdoor recreationists. Ron Arnold, one of the movement's central figures, says it is made up of people who are "alienated, frustrated and enraged" by the course of government environmental protection policies.

Multiple-use advocates rally around a core of general principles and beliefs that include a distrust of government regulation, an opposition to what they see as the "preservationist" tendencies of the environmental movement and a bedrock belief in the primacy of private property rights.

The Environmental Conservation Organization is one voice among many in this diverse movement. This passage from its 1991 action agenda crystallizes some of the movement's basic tenets:

"If there is a challenge more urgent than the protection of our environment, it is the protection of those democratic and economic principles that have guided America to become the envy of the world. Those principles are in great jeopardy largely because our national zeal to protect the environment has blinded us to the abuse we have inflicted upon our Constitution. The genius of American democracy is that ultimate authority is vested in the people. Environmental laws have proliferated in recent years because the people demanded protection of the environment. As the enforcement of those laws reveals their adverse impact upon Constitutional principles, the people, once again, rise to demand protection of their Constitutional rights.

". . . The right to own property, especially land, is the foundation of capitalism. Free enterprise is the process by which that capital is multiplied. Profit is the motive that stimulates creativity, and personal effort hones that creativity into economic gain. Any law or regulation that restricts or diminishes that process diminishes our individual and national well-being.

"Most of our current environmental legislation restricts and diminishes both the right to own property and the freedom to use that property for economic gain. . . . It is a bureaucracy of the federal government that decides what is a wetland, or what is critical habitat, or what is a scenic site, or what is historic enough to be preserved, without regard for the constitutional rights of the individual who owns the wetland or habitat or scenic or historic site. Current environmental law, for all practical purposes, confiscates these properties as 'public resources.' The owner has little say, and no effective remedy. These procedures have substantially eroded the bedrock principle of capitalism; these procedures must be corrected."

Source: *Balancing the Environmental Equation: An Agenda for a Better Future*, Environmental Conservation Organization, 1991.

land had been classified as a nontidal wetland. Describing the lot, journalist Warren Brookes noted, "When I walked through the wettest of these mostly wooded 'wetlands' last April (the wettest season), my dress shoes emerged pristinely unmuddied."

Adamski applied to the U.S. Army Corps of Engineers for the federal permit he needed to fill one-eighth of an acre of his "wetland" for the homesite. In its role as an advisor to the Corps, the U.S. Fish and Wildlife Service recommended against his permit.

Over the objections of the Fish and Wildlife Service, the Corps eventually issued Adamski's permit. But they added a stipulation that has so far prevented him from building the retirement home he had planned. The permit states that the wet-

land may be filled if Adamski compensates for the loss by constructing a one-quarter acre nontidal wetland, twice the area of the one he would fill. When the permit was issued, the cost of Adamski's retirement home skyrocketed. In addition to the actual construction costs for his house, he would have to purchase additional land and pay the construction costs of the new wetland.

For advocates of Choice Two, this story illustrates how wetland regulation can get out of hand. In the first place, they charge, it is regulatory overkill to require someone like Adamski to apply for a federal permit to fill such a small piece of land that does not fit the traditional description of a wetland.

Secondly, Adamski's story highlights the kinds of costs—both in terms of mon-

ey and delays to the landowner—that proponents of our second choice find unacceptable. In this view, wetland regulations give government the authority to confiscate private property and to interfere with legitimate economic development. When this confiscation happens without payment to landowners, the cost of conservation—clearly a public benefit—falls squarely on the shoulders of private wetland owners.

Henry Lamb, executive vice president of Land Improvement Contractors of America (LICA), voices this concern: "It is fundamentally unfair to enact legislation that forces individual landowners to bear the cost [of preserving wetlands] and to pay taxes on property that must be preserved for the public good."

Promoting Stewardship

Supporters of our second choice believe that wetland policy should be built around the idea of private stewardship. Government should encourage owners to act as stewards—or caretakers—of their land. An attitude of stewardship cannot be required by regulation. Where it does not already exist, this attitude can be fostered only through education and incentives. Incentives for landowner responsibility, these proponents say, must be based firmly in an understanding of landowner interests.

Economic concerns are among the most pressing of those interests, according to advocates of this option. The Soil and Water Conservation Society takes the position that "Private wetland owners possess an asset that may provide enormous public benefit. But wetlands nor-

Bob Hora

Property Rights in the U.S.: A Constitutional Question

While abortion-rights activists fear that a more conservative Supreme Court may soon deal a blow to their cause, an increasingly militant property-rights movement is hoping that the same ideological tilt means that its day has come.

The movement is resting those hopes on three cases, which both sides agree could have a sweeping impact on what property owners can do with their land. At issue are fundamental assumptions about the government's right to regulate land use through local ordinances, environmental legislation and other controls.

The specific cases involve a trailer park in southern California, two beach-front lots in South Carolina and a resort development planned for a remote mangrove forest in Puerto Rico. The plaintiffs are challenging, respectively, a local rent-control ordinance, a state law aimed at controlling shoreline erosion and actions taken by a Puerto Rican agency to delay a massive hotel and residential development in a proposed environmental preserve.

Although many complicated legal questions are under debate in these cases, the core issue is whether the land-use regulations in question are a "taking" under the Fifth Amendment, which states that "no person shall be deprived of . . . property without due process of law; nor shall private property be taken for public use, without just compensation."

If the court rules for the plaintiffs, governments might be required to compensate property owners for losses resulting from their inability to build or use their land as they choose. That could make governments at all levels reluctant or unable to enact or enforce laws protecting environmental preserves, wetlands, historic buildings or zoning restrictions because of the huge costs involved.

©The Washington Post, Feb. 6, 1992. Reprinted with permission.

mally yield a relatively small economic return to the owners. Faced with escalating land taxes and pressures to increase cropland for production of cash crops, too little incentive remains for private owners to retain wetland areas."

According to supporters of our second choice, wetlands will be protected when landowners, both individual and corporate, find it in their interest to do so. Therefore, proponents of this choice recommend the following:

- **Income tax incentives**. The government would offer personal or corporate income tax deductions for donations of wetlands to government or private preservation organizations, or for land exchanges that protect wetlands.

- **Property tax assessments**. State and local governments would reduce or eliminate property taxes on wetlands.

- **Direct subsidies**. Federal, provincial and state conservation reserve programs would enable landowners to receive payments for protecting their wetlands. This protection could run for a limited period of time, or it could be permanent.

- **Incentives for developing in non-wetland areas**. Governments would encourage planned development to move from wetland areas to less environmentally sensitive areas. Tools for doing this include land exchanges, systems for trading development rights and flexible zoning regulations.

- **Landowner education**. Governments would encourage stewardship through educational programs. These programs would help landowners understand how wetlands work and why they are important, as well as provide them with options for conserving wetlands on their property.

Policy tools such as these preserve landowners' ability to make choices that generate economic gain from their land. And implementing them brings into focus a question that Choice Two proponents believe should be central to all decisions about wetland policy: How much are we, as a society, willing to pay to protect wetlands? Supporters of this choice say that we must look beyond public environmental benefits and consider other values as well, such as private and societal economic well-being. Putting a price tag on protecting wetlands helps us do just that, as it helps to remove the burden of carrying out society's choices from landowners' shoulders.

Prairie CARE

Prairie CARE (**C**onservation of **A**griculture, **R**esources and the **E**nvironment) is a program that was established in 1989 by Ducks Unlimited Canada (DU)—a private, non-profit organization dedicated to preserving, restoring, developing and maintaining waterfowl habitat. The program is offered by DU; Saskatchewan Wetland Conservation Corporation; Manitoba Habitat Heritage Corporation; and Alberta Forestry, Lands and Wildlife under the North American Waterfowl Management Plan.

Prairie CARE works with farmers across Alberta, Manitoba and Saskatchewan to protect wetlands and improve the surrounding upland areas on farmlands. These three prairie provinces are in what is known as "pothole country," the defining feature of which is potholes—permanent and temporary wetlands that dot the landscape. These wetlands are among the most important waterfowl habitat in North America and important sources of water for the surrounding prairies and farmlands.

Prairie CARE recognizes that the well-being of waterfowl, successful and sustainable farming, and a healthy prairie environment all depend on wetlands being linked with well-cared-for uplands. These areas are often used by farmers for hay and other crops, and for grazing. The program aims at improving grass cover—important for sheltering nesting waterfowl—on farmlands that contain small and numerous wetlands.

Prairie CARE respects a farmer's need to make a living from these lands. Through financial incentives and technical assistance, the program encourages farmers to use their land in ways that benefit wildlife *and* the farmer. The program's major components include the following:

- Paying farmers to set aside parcels of land as natural habitat. Yearly rental payments provide the farmer with the kind of income that would be received by renting the land.

- Helping farmers change their land-use practices to provide spring and fall wildlife habitat on lands they continue to use for agriculture. Prairie CARE offers farmers financial and technical assistance.

- Purchasing lands from farmers to create intensive waterfowl management sites. Often these lands are low-productivity areas for which farmers would prefer to find alternate uses anyway.

In 1991, Prairie CARE put over 150,000 acres of farmland—belonging to some 2,200 farmers—under agreement. In addition, over 1,300 farmers participated in field demonstrations of conservation practices. And more than 17,000 acres of marginal farmland were purchased and seeded with native grasses.

A DU publication on Alberta's Prairie CARE program points out that, as in other provinces, "If a farmer has eligible lands, the financial elements of Alberta Prairie CARE can offer a steady, predictable cashflow through the 1990s and beyond. Other aspects of the program . . . demonstrate how changing technology can make conservation farming a practical alternative."

Source: Ducks Unlimited Canada, *Annual Report 1990*

Minimizing Government Control

Recent policy proposals in both the United States and Canada are designed to allow "no net loss" of wetlands or wetland functions. For advocates of Choice Two, the regulatory net cast by this kind of approach to wetland protection is too broad. They believe that a "no net loss" approach brings even the smallest bits of wetlands under government control.

According to this perspective, government jurisdiction should encompass only wetlands of critical societal importance. These are wetlands whose public environmental value clearly overrides other values. Only these key areas merit government protection, say supporters of this choice. In those few and carefully determined cases, the government must offer to buy the wetlands from the owner for a fair market price. Landowners must have the power to negotiate a price with the government and to turn down the offer should they prefer to keep their land.

A bill introduced in the U.S. House of Representatives in 1991 proposed such a program. This bill—H.R. 1330—offered a three-tiered system for classifying wetlands. "Type A" wetlands would be those with very high environmental value. A landowner (including a state or local government) whose wetlands were classified as "Type A" by the federal government would be able sell them to the government. The landowner would negotiate a fair market price, including legal fees, and would have the option to refuse to sell. Disputes over the value of the wetlands could be settled in the United States Claims Court. In most cases, the government would buy only the surface rights to the land, or water rights. The landowner would keep subsurface oil, gas or mineral rights.

Private Organizations Preferred

Still, relying on the government to preserve critical wetlands through purchasing makes many advocates of Choice Two uncomfortable. Over the long term, they believe that this approach will remove significant amounts of land from productive use and from the tax base. As they see it, this lock-up of land is based on decisions made by government bureaucracies that are insulated from the demands of the market and the will of citizens. In their view, most decisions about wetland use should be made outside the realm of government control.

Non-governmental efforts to acquire and preserve wetlands are better suited to this perspective's emphasis on private initiative and market-based choices about land use. Supporters of this option favor the preservation efforts of organizations such as The Nature Conservancy, Ducks Unlimited, the National Audubon Society and local land trusts. These organizations receive their funding from private donations. Their wetland purchases express clear choices about wetland conservation—backed by cold, hard cash.

Also, Choice Two advocates argue, we should recognize that privately owned wetlands may already be in good hands. In recent years, the concept of "corporate environmental citizenship" has gained support. Many companies are actively realizing that good environmental practices can be compatible with good business practices. For example, the Dow Corporation, a major U.S. chemical manufacturing firm, owns 840 acres (340 hectares) near the eastern Illinois city of Joliet. The land is bounded by the DesPlaines River and an Illinois state conservation area. Dow uses less than one-tenth of the land for manufacturing and maintains the rest as woodland, grassland, farmlands and wetlands. In 1981, construction of a roadway connecting Dow's new marine terminal with the main plant threatened to destroy much of the property's wetland area. But by using innovative technology, Dow designed a system of weirs to protect the new road from flooding, and to redirect the water so that new wetlands would form. Now, more than ten years later, the Dow property boasts three times its original wetland area, and offers significantly increased wildlife habitat. In addition, company efforts to prevent soil erosion from the site's farmlands and to control water quality help to ensure the health of the wetlands.

Reprinted from *The UCA News,* May 1990

This logo was designed by the Port Blakely Mill Co. for Washington State's first Wetlands Day. The company joined forces with state and federal agencies and a local school district to develop a public education program on conserving wetlands on private and public lands.

Educating for Stewardship

Proponents of this option depend upon a strong education program to help landowners make wise choices about their wetlands. For landowners and the general public alike, information about the functions and values of wetlands can help build understanding. This understanding, in turn, can create a supportive climate for responsible stewardship. And, supporters of Choice Two argue, the public needs more information about the pressures that compete with wetlands protection for a landowner's resources. As John Walter wrote in *The LICA News*, "It's telling . . . that of the two-thirds of farmers who perceive themselves to be friendly toward wildlife, more than 80 percent have never received any kind of public support (cost-sharing, tax relief, technical assistance or other forms of encouragement) for their conservation

Port Blakely Mill Co.

efforts. Yet the public expects the farmer to invest time and money in managing wildlife resources [such as wetlands] which provide little or no financial return."

From this point of view, wetland owners also need more specific information about the management options and programs available to them—and about the pluses and minuses of each. Technical assistance is also important in areas such as restoring and monitoring wetlands, using wetlands in ways that will not harm them and planning for responsible development.

Advocates of Choice Two point out that a variety of organizations and all levels of government agencies could help educate landowners. Conservation and special-interest groups, such as landowner and professional associations, could also be involved. Corporations and industry groups could get into the act, as could universities, other educational institutions and scientific research organizations.

Proponents of our second choice point out that wise stewardship of wetlands is not only in the best interest of society. It is also in the best interest of the individuals, corporations and organizations that own an important share of wetlands in the United States and Canada. Their voices, their wisdom and their rights should be central in our wetland decisions.

Wildlife habitat near rail system on Dow Chemical's land.

Laura L. Davidson/Dow Chemical

What Opponents Say

Critics of this choice argue that in the rough-and-tumble world of the marketplace, natural values will inevitably lose against more immediate economic benefits. They agree with noted naturalist Aldo Leopold, who said that "a system of conservation based solely on economic self-interest is hopelessly lopsided. It tends to ignore, and thus eventually to eliminate, many elements in the land community that lack commercial value but that are (as far as we know) essential to its healthy functioning."

Further, some critics argue that landowners—especially farmers—will not invest time and money protecting resources that provide them little financial return. For farmers, they point out, wetlands provide an important water source, especially during the summer, as well as habitat for insect-eating songbirds. In some regions, these critics say, destruction of wetlands has contributed to persistent low water tables and drought.

Other opponents maintain that this approach is motivated by an unfounded fear of government regulation. This fear, they charge, results in a sacrifice of the greater good of society. Wetland benefits extend beyond individual landowners to the society of which they are a part. These critics believe that the government is the only institution with a broad enough view to determine what's best for society.

In addition, critics maintain that many of the voluntary incentive programs proposed in Choice Two are unworkable over the long term. They do not result in a permanent commitment to preserve wetlands. And the cost to the agencies and private organizations offering the incentives can increase dramatically over time as other uses for the wetlands become more economically attractive to the land-

Private landowner on Maryland's Eastern Shore learns about wetland enhancement from a Soil Conservation Service employee.

owner. In this view, the flexibility and control that makes these programs attractive to landowners renders them insufficient for the protection of wetlands in the long run.

Finally, because government protection would extend to only a few wetlands of critical importance, opponents argue that this approach allows the continued piecemeal destruction of most of the wetland resources of the United States and Canada. One-eighth of an acre at a time, they argue, we have whittled away our wetlands until what remains in some areas is a distressingly small proportion of what was here 200-plus years ago. Proponents of Choice Two, these critics allege, would do little to prevent the same pattern from continuing.

Breaking that pattern of destruction is the first concern of people who advocate our third choice. Their proposals aim to achieve broad preservation of our remaining natural wetlands. We now turn to their perspective.

CHOICE

3

BREAKING A PATTERN OF DESTRUCTION

In 1971, the U.S. Army Corps of Engineers presented the state of Florida with what amounted to a brand new Kissimmee River. From its headwaters south of Orlando to its mouth at Lake Okeechobee, what had been 103 miles of meandering, marsh-lined waterway was now a straightened, widened and deepened 56-mile channel.

The Corps accomplished what the state of Florida had asked it to do: It made sure that killer and costly floods like those associated with a series of hurricanes in the 1940s would never again threaten the Kissimmee Basin. But in the process, it created new and potentially more serious problems.

Within five years of the project's completion, the scale of these problems was becoming evident. Straightening the river through channelization resulted in the loss of about 43,000 acres of flood plain consisting of wetlands, and opened up millions of acres of land to agricultural, housing and commercial development.

Increased pressure on the land and water, combined with destruction of the natural systems that may have been able to buffer the effects of added use, gave rise to some serious results. Game fish and waterfowl populations declined significantly. Water levels in Lake Okeechobee dropped. Drought and severe agricultural pollution created shortages of drinking water.

As early as 1976, the Florida state legislature was calling for the restoration of the Kissimmee and its surrounding wetlands. Now, Florida and the federal government are considering a massive public works project. Over the course of 15 years and to the tune of hundreds of millions of dollars, the project would restore the river to its original channel.

Proponents of our third choice see the story of the Kissimmee River repeated in big ways and small—and with alarming frequency—across the United States and Canada. They believe that all too often, in our zeal to control our surroundings, we harm the natural systems that sustain us and all other life on earth.

The Kissimmee River and flood plain prior to (left) and after (right) channelization.

Advocates of Choice Three maintain that the destruction of a huge proportion of our wetlands now leaves us little room for error in our policies. They call for a cautious policy based on preserving our remaining wetlands and on reversing the attitudes that have led us to this precarious point. David Whitesell, former executive vice president of Ducks Unlimited, voiced this commitment to preserving wetlands: "We have talked and compromised our way through the loss of 100 million acres of wetlands, and there is no longer room for any more draws or lost fights."

Kissimmee River restoration model.

Playing for Keeps

Several years ago, I agreed to serve on Illinois' Lake County Environmental Health Advisory Committee, a decision I've not regretted making. Having spent nearly 40 years in the professional fish and wildlife business—over half of it with Ducks Unlimited in the international conservation arena—the local meetings seemed at first to be low-keyed and not terribly controversial. But all this changed quickly one afternoon in early March when a young biologist presented an environmental impact statement on a proposed housing development in Lake County. A 100-acre wetland was at stake, and the young man reported that the developer could manage to save all but 15 acres of the marsh.

What irritated me most was the fact that the biologist, an employee of an environmentally oriented agency entrusted with the safeguarding of natural resources, seemed pleased with the report. Most of the others involved also regarded the piece of business as one in which a satisfactory compromise had been made, and they were prepared to move on to

Dale Whitesell

other business when I decided to tell them what I thought of the 15-acre loss.

The situation had actually taken me back to my days with the Ohio Division of Wildlife, an outfit that first offered me employment back in 1950. The following year I traveled to my first North American Wildlife Conference which was held in Miami. A hard-nosed old character, who had spent 40 years with the Division, made a comment to me which has over the years made an increasing amount of sense. Having attended the North American [conference] regularly during the course of his career, he had determined that the same conservation issues were

addressed year in and year out. We spend more time discussing them than trying to win them, he said. Each year, we lose yardage, and in many instances there is not a lot of game time left.

Wetland loss is a case in point. In the 1700s there were an estimated 216 million wetland acres in the United States. Today, there are at best 92 million acres remaining, and each year we continue to compromise and pitifully justify the loss of nearly one-half million more. With approximately 1,200 wetland acres slipping away each day, I'd decided during the meeting that our 15 wetland acres were simply not going to be part of that total pot. . . . There is no compromise . . . when it comes to altering wetland habitat. Once it is drained, you cannot replace it, and what you have lost involves something more than what was once home to a delicate and complex system of marsh life.

Source: *Ducks Unlimited*, editorial by Dale E. Whitesell, former executive vice president of DU, May-June 1987

Don't Break It if You Can't Fix It

Under Choice Three, the first order of business should be to protect the natural wetlands we have remaining. We simply do not know enough to do otherwise. Proponents point out that even the most skilled experts in creating or restoring wetlands cannot completely duplicate natural wetlands. Most efforts at restoration or creation, they say, do not even come close to substituting for a real wetland.

Advocates of this third choice believe that our knowledge of wetlands and their role in the web of life on earth is only in its beginning stages. They find it striking and somewhat humbling that each discovery in our understanding of wetlands—and of nature in general—opens up a whole new area of questioning. No matter how much we learn, it seems there is always more to know.

Choice Three proponents believe that what we *do* know about wetlands suggests that they are masterpieces of natural design. They are intricately balanced, teeming with life, varied and dynamic. And they are a critical part of nature as we know it.

Advocates of Choice Three underscore the wide range of wetland functions that we know about—from maintaining water quality to stabilizing the atmosphere, and from supporting a variety of plants and animals to recharging the human spirit. In this view, preserving wetlands in their natural complexity is the only sure way to maintain all of these functions. And it is the only way to ensure that future generations will share in the benefits of our diverse and evolving natural world.

Supporters of this option believe that more and more people are starting to see the value of wetland preservation. Talking about those changes in public attitude, Doug Hagen of the Ontario Ministry of Natural Resources Wildlife Branch said, "When we first began Ontario's program of identifying and protecting wetlands of provincial importance, the idea we worked with was 'compatible use.' We tried to stop the most destructive uses of these areas. But activities that weren't as harmful were allowed. Now, our message is more like, 'Stay off provincially important wetlands.' The public has demanded this."

Preservation with a Twist

Advocates of Choice Three propose a wetlands policy that admits the limitations of our knowledge of nature and of our ability to manipulate nature to our long-term benefit. That policy begins with wilderness-level preservation of natural wetlands. Supporters of this option believe that the government should buy up wetlands on a large scale. These purchases should begin with the largest, most biologically significant and most threatened wetlands.

Choice Three proposes that these wetlands, as well as those the government already owns, be permanently designated and managed as wilderness areas. In these areas, nature should run its course and the land should be protected from damaging human use. Most wilderness areas are protected by excluding development and many other human activities. Advocates of Choice Three believe that these traditional wilderness-protection methods may not be enough to ensure the permanent protection of these wetland wilderness areas.

Biosphere Reserve: An Idea in Action

Along the Atlantic coast of Maryland and Virginia lies a chain of barrier islands more than 100 miles long. These islands are the centerpiece of the Virginia Coast Reserve: 40,000 acres of islands, saltmarsh tracts and adjacent mainland sites. The Reserve is owned and managed by The Nature Conservancy, an international membership organization committed to the preservation of natural diversity.

Over the past twenty years, The Nature Conservancy has successfully protected these islands and marshes from development, establishing them as a nature preserve. But the long-term protection of the island ecosystem—the complex web of life that exists on and around the islands— is uncertain. The salt marshes, which separate the islands from the mainland, are the critical component of this ecosystem. They are threatened by increasing residential, resort and retirement development taking place outside the boundaries of the sanctuary. As an example, a Nature Conservancy publication explains, "A high density retirement development far removed from the barrier island sanctuary might still significantly alter the productivity of the island ecosystem by introducing sewage, lawn nutrients, pesticide residues and petroleum products into the watershed, which ultimately deposits them into a seaside bay."

Threats like these led The Nature Conservancy to realize that a more comprehensive and far-reaching plan for ensuring the health of the island ecosystem would be needed over the long haul. Thus, they began working with the "biosphere reserve" idea.

Curtis J. Badger explains how this idea works in his article, "Eastern Shore Gold." A biosphere reserve "has a core

Visitors to the Virginia Coast Reserve.

area that receives maximum protection, a buffer zone that is managed to protect the core and a transition area, or 'zone of cooperation.' This last area is managed with private landowners for business, recreation and traditional uses that are in keeping with the biosphere reserve purpose.

"The Virginia Coast Reserve provides a ready example. The barrier island, coastal bay and saltmarsh ecosystem constitutes the core area; the adjacent mainland and watershed form the buffer zone, where traditional uses such as sustainable seafood harvesting and farming are encouraged; and the transition area includes farms, village clusters, businesses, recreation facilities and light, compatible industry. The land in both the buffer and transition zones is both privately and publicly held, and the uses are governed by conservation easements and deed restrictions, and by local zoning and subdivision ordinances."

The Nature Conservancy is working in partnership with local, state and federal governments; businesses; farmers; watermen; developers; civic groups and private conservation organizations to ensure the fu-

ture protection of the preserve. Federal and state agencies, for example, are cooperating by establishing new parks and refuges around the reserve. Private landowners are working with Conservancy staff to draw up conservation plans. Local governments are passing ordinances to manage growth. These activities help to direct much of the development away from vulnerable wetlands and waterways.

These efforts, as Badger points out, are "more than a sophisticated land-use planning technique. [They address] cultural values, history, and economic development, and [affect] the entire population of the region in one way or another." These partnerships are helping people keep the way of life they love by preserving traditional industries such as seafood and sport fishing, as well as traditional values, land-use practices and community vitality that are threatened by escalating property values and the push for development.

Sources: *The Islands*, Fall 1989; and *Nature Conservancy*, Curtis J. Badger, July/Aug. 1990

More than any other ecosystem, proponents of this view maintain, wetlands are extremely sensitive to changes in surrounding areas. Even natural preserves as large as the Everglades are susceptible to outside pressures that compete for—and pollute—the water that feeds them.

So advocates of Choice Three favor a "biosphere reserve" or "bioregion" approach to preservation. Under this approach, a core area of wilderness is surrounded by a buffer zone in which only ecologically sensitive development is allowed. In other words, the long-term health of a protected wetland area would be maintained by tightly controlling the use of the neighboring uplands and water sources.

Other Protection Strategies

Proponents of Choice Three recognize that all wetlands cannot be wilderness areas. So they also rely on other means of ensuring permanent protection. Here are some of the methods they propose:

- **Perpetual conservation easements** through which government or nonprofit organizations pay landowners to permanently preserve their wetlands. If the land is resold, these restrictions on its use are attached to the title.

- **Matching funds** for acquiring and setting aside wetlands of local importance. The Reinvest In Minnesota program, for example, matched funds raised by a group of local businesses, conservation groups and citizens to buy what is now the A Shau Valley Wildlife Area.

- **Nonprofit wetlands preservation trusts** that would acquire wetlands and surrounding areas by donation, purchase or land exchange. The government would encourage landowners to participate through tax benefits and other forms of technical and financial assistance.

- **International agreements** that commit governments and private organizations to wetland preservation.

Regulating Private Wetlands

We all may wish to live in a society in which people willingly protect the natural environment. However, Choice Three advocates believe it is dangerous to assume that we do live in such a society. Proponents of this choice believe that government regulation of privately held wetlands is an important part of a policy aimed at preservation. They point out that a high percentage of the population—and private land ownership—in our countries tends to be concentrated near the wetter regions. So the pressure to develop in those areas is strong.

Proponents of this view believe that the government is well within its power to restrict development in environmentally sensitive and critical areas such as wetlands. In fact, this control is part of the government's duty to protect societal well-being.

From this perspective, government regulation should aim to ensure that we do not—even inadvertently—whittle away at wetlands when they are such an important part of the earth's life support system. It is easy, supporters of this option believe, to ignore small wetland areas that do not seem very impressive. But even small areas are ecologically important and need to be protected by government regulation. And wetlands that are wet for only part of the year need that same protection. Timothy Searchinger, an attorney with the Environmental Defense Fund, writes that "some find it hard to understand that an area that is not under water much of the year could really be a 'wetland.' The President himself . . . suggested that [we should only] preserve areas that were 'duck wet.' Wetlands, however, are valu-

Mark Müller/Iowa State University of Science and Technology

Lisa Wilcox

able in large part because they *vary* in wetness. . . . Ducks, no less than other animals, rely heavily on lands that are wet on the surface for only brief periods."

For supporters of this option, mitigation has no place in a regulatory program. They believe that the goal of wetland regulations should be to allow as little loss of wetlands as possible. Having a mitigation program in place, they argue, contributes to the view that destroying wetlands is acceptable, because we can make up for the losses. At least until we know much more about the intricate workings of wetlands and how to replicate them, advocates of Choice Three find the idea of mitigation to be a dangerous illusion. Allowing natural wetlands to be replaced by inferior artificial wetlands gradually decreases the health of the environment.

A Broad View of the Value of Wetlands

Although many opponents of wilderness and land-use regulation think otherwise, advocates of Choice Three believe that preserving wetlands does not lock up resources from human use. Rather, keeping these unique and vital natural systems intact assures us a continuing opportunity for research and study. In this view, studying wetlands will help us better understand nature and our place in it. Research may lead to greater knowledge of basic environmental processes. And it may help us address worldwide environmental problems such as global warming, and local problems like water pollution.

Wetlands almost certainly harbor resources yet to be discovered, such as plants whose medicinal value is unknown. About half of all of today's medicines have key ingredients taken from plants. Supporters of Choice Three caution that, in destroying wetlands, we also destroy the plant life that is unique to them. By destroying that plant life, we may eliminate options—the key to curing cancer or AIDS, for example—that future generations could find useful or essential to their survival.

Proponents of Choice Three also believe that research may improve our ability to restore and create wetlands. But they caution that we may never know enough to ably manage the natural order. From their perspective, past and current policies overemphasize the use of wetlands for demonstrable human benefits. This human-centered approach has led to a skewing of priorities. We have traded a healthy and diverse environment for one that is less healthy, less diverse and less able to withstand the pressures we put on it.

Advocates of this choice read trends such as a 30-year decline in U.S. and Canadian waterfowl populations as signs of worsening environmental health. Loss of wetland habitat has been a major factor in the decline of duck species such as the mallard, northern pintail and blue-winged teal. From this perspective, the dwindling populations are a highly visible warning about the impact of our actions.

Choice Three supporters believe there is a very practical reason for being concerned about decreasing duck populations: The same environment that supports ducks supports humans. One day, supporters of this choice warn, we will have to lie in this bed we are making.

Angelica atropurpurea: A close European relative of this U.S. wetland plant was used in folk medicine to treat respiratory ailments and to stimulate the appetite.

A Spiritual Awakening

Proponents of our third choice believe that we can make a different and more comfortable bed for ourselves if we change our basic approach to wetland policy. They support policies that are less human-centered and more concerned with nature's overall balance.

This kind of policy will allow us to act as a part of the whole, rather than to try constantly to dominate the rest of the whole. In this more cooperative mode, advocates of Choice Three believe that we will find relief from the endless demands of being the overlord and manager of nature. And we will find a great spiritual satisfaction that we miss in our essentially selfish view of the world. Recognizing our connection with the rest of nature will lead us to be less destructive and to appreciate more fully nature's beauty, intricacy and inherent worth.

Advocates of Choice Three seek a change that goes beyond wetland policies. But in this quest for the spiritual awakening that will help get our societies on the right track, what we do with wetlands can play an important role.

In this view, wetland preserves can provide the direct contact with nature that might spark this awakening and feed our understanding. They should become educational centers where adults and children alike can learn the ways of nature. And, Choice Three proponents encourage recreational use when compatible with the health of wetland areas. To them, recreation is valuable because it brings people into close contact with the natural world. Activities such as canoeing through quiet marsh waters, figuring out the difference between a redhead and a canvasback duck, capturing the vitality of wetlands on film or learning the waters well enough to find and hook a bass deepen our understanding of nature, and our connection to it.

Restoring Wetland Areas

Preserving existing wetland areas takes clear precedence in the policy proposals outlined in this choice. But supporters of this option also believe that it is important to restore areas that were once wetlands. They acknowledge that the damage done to these areas cannot be completely repaired. But to them, restoration offers a means of rebuilding our stock of wetlands. And it allows us to play a part in healing the earth, rather than destroying it.

In this view, wetlands that were drained or altered for agricultural use may be prime candidates for restoration. Many of these lands, now used for crops or pasture, would be relatively easy to restore, especially compared to former wetlands that now have buildings or blacktop on them.

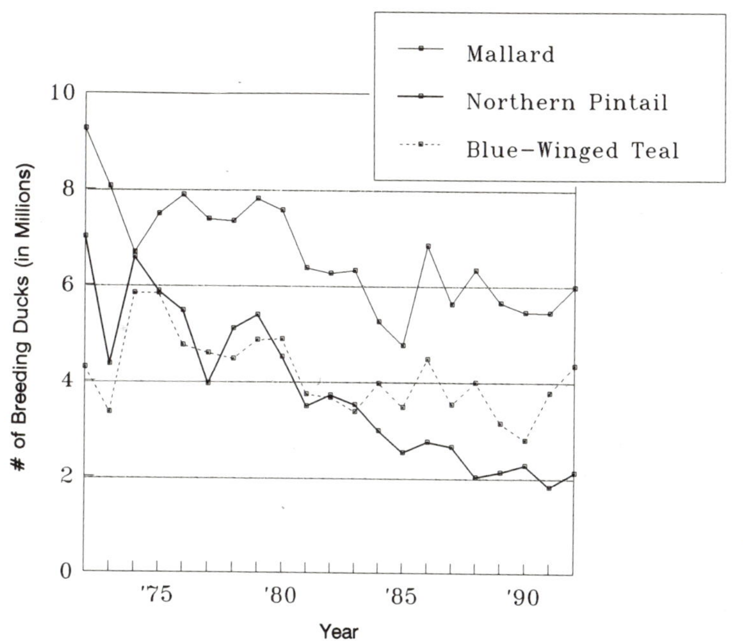

Some wildlife biologists believe that the decrease in duck populations is caused by diminishing wetland habitat.

Advocates of Choice Three propose permanent wetland reserve programs. Owners of certain kinds of altered wetlands could sell the land to the government or to a nonprofit organization. Or they could put it under a permanent conservation easement. The government would pay for at least part of the cost of restoring the wetlands.

Urban wetlands are also high on the Choice Three priority list for restoration. "Those soggy spots may be degraded and neglected, their edges a tangle of trash and old tires. But their potential for recreation and environmental education may make them some of the most essential wetlands in the nation," explains Oregon writer Phillip Johnson. "The great majority of Americans will come to know and care about wetlands only through the treks they make to metropolitan marshes."

What is critical from this perspective—and what sets it apart from Choice One's approach to restoration—is that restoring wetlands should be used only as a means of revitalizing areas that have already been damaged. We should rely upon our limited ability to restore wetlands only as a healing measure. It must not be used to justify more destruction.

In the eyes of Choice Three supporters, permanently preserving wetlands is the best way to ensure their future health and survival—and ours, too. As Canadian environmental reporter Michael Keating suggests, wetland destruction continues "thanks to a pioneering attitude yet to be tamed by a realization that wetlands are a very important and valuable part of the landscape." Advocates of Choice Three believe that our zeal for development and control must give way to a more cooperative attitude. And that we must stop destroying wetlands.

What Opponents Say

Opponents of this approach to wetland policy point to its high economic cost. Taken together, government outlays for wetland purchase, management, regulation and research total up to a hefty amount. Critics maintain that taxpayers will be unwilling—and even unable—to pay this price.

The high cost of government action is merely one cost of this approach. Oppo-

Kids coming into close contact with nature.

Tamarack Camps

An agricultural wetland before (left) and after (right) restoration.

nents argue that Choice Three requires people to absorb an additional cost: that of lost economic opportunity. This approach to wetland management, they charge, locks up valuable land and resources from most economically beneficial uses. And it ignores the need and the right of people to support themselves. Wetlands are a resource like all others, to be used to benefit humanity, not preserved in a bubble.

For some opponents, this choice amounts to an imposition of the will of a fairly small group of well-off environmentalists on the rest of the public. They are fed up with what they view as "environmentalist land grabs." People like Grant Gerber, founder of the Wilderness Impact Research Foundation, allege that farmers, miners, loggers and others who "use the land" have a common enemy that is attempting to eliminate their way of life: "A tiny group of idealogues that are backpackers."

Finally, critics charge, these wetland policy proposals are based on a pie-in-the-sky notion of harmonious coexistence with nature. It is impossible, they say, to practice environmental non-intervention. And it is unrealistic to believe that, without proper management, wetland areas would continue to exist and to provide the benefits that we value. In the first place, they argue, wetlands naturally fill in and become dry land over time. Further, these critics believe that humans have had too much impact on natural systems already to back off, let nature take its course and live with a much more modest effect on the environment. Political writer Walter Truett Anderson puts the idea this way: "The American continent has been transformed; it is now an artificial ecosystem and it must be managed by human action. This cannot be stopped, now, nor can we return to a natural order untouched by human society. We are at the controls, whether we like it or not."

Discovery at a wetland.

WETLAND OPTIONS: FINDING A WAY OUT OF THE MIRE

We may not agree that humans are irreversibly at the environmental controls. But we *are* at the controls in a sense. As citizens, we have the opportunity to take the helm in directing the course of wetland policy. In Canada and the United States, public debate continues over alternative approaches to making decisions about wetlands. Discussions you have with your fellow citizens can help to chart a direction for these policies.

The task facing us is to take a hard look at our options, and to consider the positive and the negative consequences of each approach.

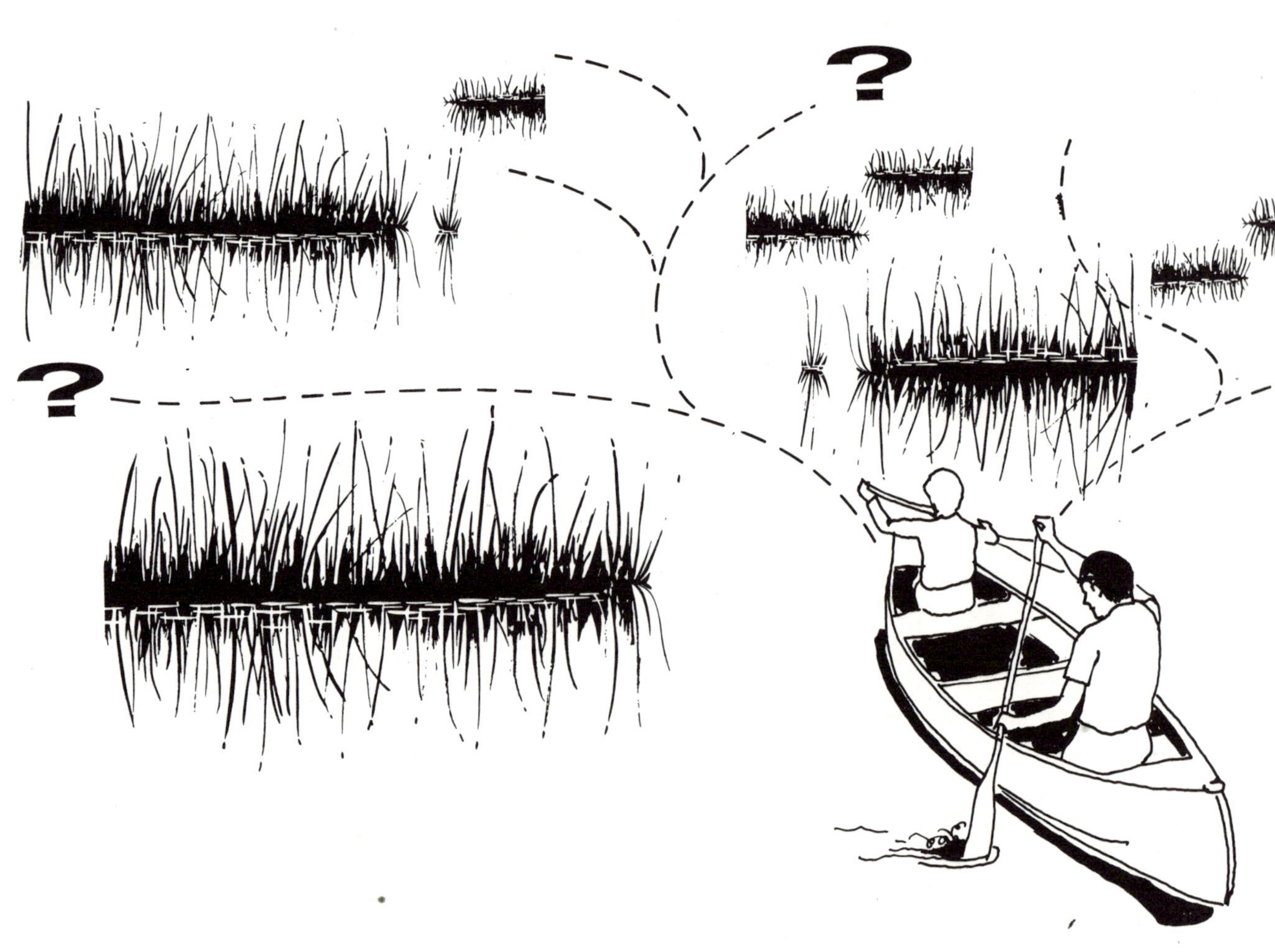

Courses of Action

As our first choice suggests, wetland policies could focus on the benefits that wetlands provide us. Supporters of this option want decisions about wetland use to be based on a detailed inventory of wetlands and what they do. With that knowledge, they believe that we will be able to determine the most societally beneficial use of each wetland. Choice One proponents believe that most decisions about wetlands should be made locally. At the local level, they say, the environmental, economic and social conditions particular to each situation would come into focus.

Alternatively, we could adopt a strategy consistent with our second choice. Proponents of this approach put forth policies that rely on the stewardship of private landowners. They want to establish a climate of support for the efforts that wetland owners make to take care of their land. They believe it is unfair—and to our countries' economic disadvantage—to ask landowners to shoulder the cost of protecting wetlands. Society as a whole benefits when wetlands are protected. And society should help foot the bill. In this view, the most appropriate policies center on voluntary conservation, economic incentives, landowner education and the efforts of non-governmental organizations.

Or, as advocates of our third choice propose, we could make it our first order of business to protect the natural wetlands we have remaining. To break what they see as a pattern of destruction and domination, proponents of this approach suggest policies that are more respectful of nature. The program they propose relies on permanent means of preserving wetlands, such as government acquisition and perpetual conservation easements. It depends, as well, on strong regulation of privately owned wetlands. In addition, advocates of this choice believe that a change in our basic attitude toward nature is needed. They would promote this change through education, nature experiences and the healing activity of wetlands restoration.

Productive Harmony

On January 1, 1970, U.S. President Richard Nixon signed into law the National Environmental Policy Act. The law was intended to institutionalize a new environmental consciousness. In its declaration, Congress made it the goal of the federal government "to create and maintain conditions under which man and nature can exist in productive harmony."

More than 20 years later, there is still plenty of disagreement over what those words mean. Nowhere is that disagreement clearer than in the debate over wetland policy. In the United States and Canada alike, it is a debate that challenges us to address some tough questions: How should we balance individual rights against societal obligations? What are legitimate roles for government to play? What orientation to nature should lie at the heart of our policies? To economic development? If we are to make progress in the debate over wetland policy, we must move toward a notion of "productive harmony" that we can all accept.

In the end, it is this need for common ground that brings us together as citizens to talk about our wetland policy options. In these discussions, we may begin to build a policy framework that reflects our best thinking about what to do with our wetlands.

ACTION AND POLICY CHART

These two pages display a chart listing various ways in which individuals, communities, businesses, industries and government bodies can take action to deal effectively with the wetland issue. The chart is organized according to the three different choices presented in this book.

CHOICES

#1

Based on thorough knowledge of our wetlands and their value, the government can decide how different wetlands should be used. This approach ensures that a diversity of societal benefits will be gained from wetlands.

#2

Landowners should have the final word in decisions about wetland use. Government should use incentives and education to encourage responsible private stewardship. This approach offers a conservation strategy consistent with preserving economic vitality.

#3

Through government acquisition and regulation, we should permanently preserve most remaining wetlands. Through education and direct contact with wetlands, we should promote a less-domineering relationship with nature. This approach offers a prudent and spiritually satisfying response to the enormous historical and continuing destruction of wetlands.

INDIVIDUAL ACTIONS

#1
- Comply with government requirements for wetland development permits.
- Participate in public hearings and community forums about specific wetland decisions.
- Cooperate with government efforts to inventory wetlands by providing information about, and access to, privately owned wetlands.

#2
- Support nongovernmental organizations that purchase and manage wetlands, or that provide support to landowners for wetland conservation.
- Make wise decisions about the use of privately owned wetlands.
- Take advantage of government education and incentive programs.

#3
- Preserve privately owned wetlands.
- Support nongovernmental organizations that purchase and permanently preserve wetlands, and that implement the "biosphere reserve" management approach.
- Cooperate with government and private efforts to protect wetlands from harmful activities outside their boundaries.
- Participate in wetland education programs and take advantage of recreational opportunities in wetland areas.
- Contribute to efforts to restore local wetland areas.

COMMUNITY ACTIONS

#1
- Consider the costs and benefits of various management or development options for local wetland areas.
- Encourage innovative uses of wetlands, such as Arcata's use of wetlands in sewage treatment.
- Disseminate information about local wetlands, conflicts or pending decisions about their use, and relevant government policies.
- Provide public forums for the discussion of local wetland-related issues.

#2
- Provide education about wetlands and related topics to landowners and the general public.
- Establish organizations that purchase and preserve locally significant wetlands, as well as channel public support for such activities.

#3
- Establish nature centers at local wetlands to provide education and contact with the natural world of the wetlands.
- Establish local fund-raising efforts and groups aimed at taking advantage of government support for acquiring and permanently preserving local wetlands.
- Channel public energy into efforts to restore damaged wetlands.

BUSINESS AND INDUSTRY ACTIONS

- Invest in research and development of processes that utilize wetlands for functions such as municipal or mining waste treatment.
- Continue to develop new uses for wetlands and their products, such as peat-fueled energy production and medicinal products.
- Utilize conservation technologies that allow development to coexist with sustained wetland functions.
- Invest in research and development of techniques for wetland creation, enhancement and restoration.
- Comply with government requirements for wetland development permits.
- Offset damage to wetlands caused by business activity through creation or restoration of other wetland areas.

- Make responsible decisions about the use of privately owned wetlands.
- Utilize innovative methods and technologies to integrate wetland conservation with business operations.
- Cooperate with government or privately sponsored programs aimed at protecting wetlands.
- Work with local communities and schools to further public knowledge about wetlands.

- Cooperate with government efforts to purchase important wetlands.
- Comply with government regulatory programs.
- Establish or support local wetland education centers and programs on company-owned wetlands.
- Participate as corporate citizens in local efforts at wetland preservation.
- Restore wetland areas damaged by corporate activity.

STATE/PROVINCIAL AND LOCAL POLICIES

- Collaborate with national level government scientists and officials in conducting wetland inventories and setting standards for classifying wetlands.
- Provide forums for public input into decisions about the use of nationally regulated wetlands.
- Design and implement land-use regulations that ensure balanced and reasoned decision-making about wetland use.
- Require "mitigation" as a condition for impairing important ecological functions of wetlands.
- Disseminate information and research results to policymakers, wetland managers, landowners and the general public.

- Reduce or eliminate property taxes on wetlands.
- Provide landowner education programs and income tax incentives consistent with national programs.
- Establish conservation reserve programs that pay landowners for protecting their wetlands.
- Encourage planned development to move away from wetland areas through land exchanges, systems for trading development rights, and flexible zoning regulations.

- Pay landowners to permanently preserve their wetlands through perpetual conservation easements.
- Provide matching funds for acquiring and permanently preserving wetlands of local importance.
- Establish nonprofit wetland preservation trusts capable of acquiring wetlands and the surrounding areas by donation, purchase or land exchange. Provide tax benefits and other forms of financial and technical assistance to participants.

NATIONAL GOVERNMENT POLICIES

- Conduct national inventories of wetlands that identify the natural values of wetlands and the health of the resource in different areas.
- Preserve wetlands of national significance through government purchase and landowner incentives.
- Allow unrestricted use of wetlands of comparatively low natural value.
- Regulate the development of all other wetlands, ensuring that decisions about their use are matters of public decision-making.
- Provide technical assistance to states, provinces, and localities to help them comply with national regulatory guidelines.
- Fund research into wetland creation and restoration, and into understanding the basic functioning of natural wetlands.
- Target research funds at evaluating current techniques for protecting and managing wetlands.

- Avoid violating private property rights and entrepreneurial initiatives.
- Foster an attitude of private stewardship of wetlands through landowner education programs aimed at providing basic information about the natural values of wetlands and options for conservation.
- Offer personal or corporate income tax deductions for donations of wetlands to government or private preservation organizations, or for land exchanges that protect wetlands.
- Establish conservation reserve programs that pay landowners for protecting their wetlands.
- Purchase wetlands of extremely high natural value from private landowners who are willing to sell.

- Purchase natural wetlands, beginning with the largest, most biologically significant and most threatened.
- Provide wilderness-level protection for and intensive control of ecologically harmful activity in areas surrounding all wetlands under government ownership.
- Enter into international agreements that commit the country to wetland preservation.
- Provide tax benefits and other forms of financial and technical assistance to landowners who participate in various programs to permanently preserve wetlands.
- Strongly regulate all wetlands.
- Fund research into how wetlands naturally function.

ENVIRONMENTAL ISSUES FORUMS

The Environmental Issues Forums (EIF) program brings citizens together in locally initiated, non partisan discussions about environmental problems that concern them. EIF is based on the enormously successful National Issues Forums model, an update and re-creation of the traditional town meeting, pioneered by the Kettering Foundation.

Civic and education organizations—high schools and colleges, libraries, service organizations, religious groups, government agencies and many other types of groups—convene forums and study circles in their communities as part of the EIF program. Each participating organization assumes ownership of the program, adapting the approach and materials to its own mission and to the needs of the local community.

Here are answers to some of the most frequently asked questions about EIF:

WHAT HAPPENS IN FORUMS AND STUDY CIRCLES?

EIF forums and study circles aim to stimulate and sustain a certain kind of conversation—a genuinely useful conversation that moves beyond the bounds of partisan politics and the airing of grievances to mutually acceptable responses to common problems. Unlike most public discussions or debates about environmental issues, forums and study circles invite discussion about each of several approaches to an issue, and encourage in-depth consideration of the costs, benefits and consequences of those approaches in light of individual and community values. The discussions do not aim to develop a specific group consensus; rather, they aim at developing a common sense of community problems and priorities.

CAN I PARTICIPATE IF I'M NOT WELL INFORMED ABOUT THE ISSUES?

People don't need to be experts to discuss environmental issues. The central task of EIF is not to help participants acquire a detailed knowledge of environmental issues, but to focus on what public actions should be taken. That's a matter of judgment that requires collective deliberation. Forums and study circles help people grasp the dilemma at the heart of each issue, and to understand basic facts and trends. And, most important, they help participants ponder and discuss the kernel of conviction on which each alternative policy approach is based, to sort out conflicting principles and preferences, to uncover areas of disagreement and to work toward areas of agreement.

DO FORUMS AND STUDY CIRCLES LEAD TO POLITICAL ACTION?

Neither the organizations that run forums and study circles locally nor the North American Association for Environmental Education (NAAEE) advocates partisan positions or specific solutions to environmental problems. The purpose of forums and study circles is to influence the process of addressing environmental issues in a more fundamental way. Before elected officials decide upon specific proposals, they need to know what kinds of initiatives the public favors. Forums and study circles provide an occasion for people to consider alternative viewpoints, and move toward a decision on what broad direction public action should take.

WHAT DO I DO IF I WANT TO CONDUCT AN EIF FORUM OR STUDY CIRCLE?

NAAEE offers a training program for individuals interested in learning more about EIF and how to run forums and study circles. There is also a growing network of NAAEE professionals who have conducted forums and may be able to help in your community. In addition, the Kettering Foundation has a great deal of information available on how to plan and conduct forums and study circles through the National Issues Forums program; most of these materials can be easily adapted to the EIF program.

For information on obtaining materials to organize and lead forums, or to learn about training opportunities, contact the EIF Coordinator at NAAEE, Suite 400, 1255 23rd St. NW, Washington, DC 20037. Phone (202) 467-8753, fax (202) 862-1947.

For information on how to order additional books in the EIF series, contact the NAAEE Publications and Member Services Office at P.O. Box 400, Troy, OH 45373. Phone and fax (513) 676-2514.